Statistical Physics of Condensed Matter Systems

A primer

Online at: https://doi.org/10.1088/978-0-7503-2269-0

Statistical Physics of Condensed Matter Systems

A primer

Luciano Colombo

Department of Physics, University of Cagliari, Cagliari, Italy

IOP Publishing, Bristol, UK

ISBN   978-0-7503-2269-0 (ebook)
ISBN   978-0-7503-2266-9 (print)
ISBN   978-0-7503-2267-6 (myPrint)
ISBN   978-0-7503-2268-3 (mobi)

DOI   10.1088/978-0-7503-2269-0

Version: 20221101

IOP ebooks

British Library Cataloguing-in-Publication Data: A catalogue record for this book is available from the British Library.

Published by IOP Publishing, wholly owned by The Institute of Physics, London

IOP Publishing, No.2 The Distillery, Glassfields, Avon Street, Bristol, BS2 0GR, UK

US Office: IOP Publishing, Inc., 190 North Independence Mall West, Suite 601, Philadelphia, PA 19106, USA

Contents

Part II Quantum statistical physics

3 The statistical description of a quantum system — 3-1

4 Thermal properties of quantum gases — 4-1

5 Other quantum systems and phenomena — 5-1

Foreword

The creation of a common European space for higher education, where students can easily train in a network of different European academic institutions has and continues to require new textbooks, separating the basics concepts taught in the first three years from the more complex and technical applications that characterize the last two years. These three volumes, primers in atomic and molecular physics, in solid-state physics, and in statistical mechanics greatly contribute to this need. As well explained by the author in his preface, these three volumes arise from the desire to expose students, even those who will leave their studies after the first three years, not so much to the casuistry of applications but to the beauty of the fundamental principles ruling the structure of matter. Forging thinking, rather than providing notions, is the spirit fuelling this editorial project.

It is not easy to write a primer on a subject as important and as vast as the statistical description of matter. It takes considerable effort to decide which topics to touch on and how to condense them into a clear and concise form, while maintaining the organic nature of the project. Luciano Colombo limits himself mostly to the description of non-interacting particles, whether classical or quantum, although a desire to go further transpires here and there in the text. This choice allows him to focus on the foundations of statistical mechanics, on how to *partition* particles among the energy levels, on how the seemingly simple concept of equal probability of microstates is transformed through the writing of a partition function into a powerful tool for predictions at the macroscopic level.

The primer in statistical mechanics is composed of five chapters. The first discusses the basic ideas defining statistical mechanics, the concept of ensemble, the correspondence between the average over time and the average over microstates, the process of equilibration as the search for the distribution that maximizes the appropriate probability conditioned to the proper constraints, for example, the conservation of the number of particles or the energy of the system. Special attention is given to the concept of temperature and entropy and how entropy teaches us about our ignorance of the microscopic state of systems. By the end of the chapter, the student has definitely understood that statistical mechanics offers a quite elegant microscopic explanation of thermodynamics and its axioms.

In chapter 2, the student learns how to build on the foundation just laid and how to predict the behaviour of systems of non-interacting particles (either atoms or molecules), even in the presence of electric and magnetic fields. Here, too, Luciano prefers to skip the commonly used description based on momenta and generalized coordinates, directly focussing on the alternative description based on energy levels, elegantly showing how the quantization of energy reveals its presence in the progressive unfreezing of rotational and vibrational degrees of freedom.

In chapter 3, Luciano introduces the quantum features associated with the particle wave function symmetry. You can perceive the amazement that students must feel as they realize that the simple imposition of the wave function particle exchange symmetry completely revolutionizes what had been learned for systems of distinguishable particles. Attraction or repulsion between particles arises as a consequence of quantum indistinguishability! Novel important concepts as the Fermi energy and the ground state energy are introduced and their physical role explained. The effect of quantum features on the temperature dependence of thermodynamic quantities and the difference with respect to the classical ideal gas emerge clearly and comprehensively.

Chapter 4 plays a role analogous to chapter 2. The power of the formalism learned in the previous chapter is shown in action, this time for quantum statistics. Here the temperature dependence of the properties of a free electron and phonon gases are discussed in detail and compared with experimentally measured trends. A short chapter 5 extends the fourth to two very important topics, black-body radiation and Bose–Einstein condensation.

I alluded that Luciano had to make a difficult choice in selecting the topics in this Primer. As I also said, despite the focus on non-interacting particles, here and there in the text, some important physics originating by the interaction between atoms and molecules peeps out. The effect of turning on attraction between atoms in classical gases is treated as a small interlude that stimulates and broadens the reader's horizon. Of course, these small inclusions cannot do justice to the many fields in which statistical mechanics has found remarkable applications in recent decades. It is nice, therefore, to find at the end of the Primer a list (chapter 6) of 'classical' topics not previously covered. Equally important is the short list of fields where statistical mechanics has played or is playing a fundamental role (critical phenomena, complex systems, non-equilibrium).

Finally, I want to emphasize the significant effort that has been made not only in the selection of the material but also in the way the material is presented. Each chapter is opened by a short syllabus which very clearly outlines the chapter content. The text is accompanied by a few figures, particularly impressive and well annotated, certainly helpful in conveying their message. Figures are also used to explain experimental techniques allowing scientists to measure the calculated quantities or to compare theoretical predictions with experimental results. Derivations are carried out step by step. The student is never left with that bitter sense of incomprehension engendered by missing steps. When the mathematical load is excessive, it is not suppressed but moved to the different appendices. This simplifies the reading of the text but at the same time offers the possibility of enjoying the continuity of the logical/mathematical process.

In short, an excellent text for students at their first encounter with statistical mechanics, which I am sure it will be adopted in several physics courses. Thank you Luciano for writing it.

Francesco Sciortino
Department of Physics
'Sapienza', University of Rome
July 2022

Presentation of the 'Primer series'

This is the third volume of a series of three books that, as a whole, account for an introduction to the huge field usually referred to as 'condensed matter physics': they are respectively addressed to atomic and molecular physics, to solid-state physics, and to statistical methods for the description of classical or quantum ensembles of particles. They are based on my 20-year experience of teaching undergraduate courses on these topics for bachelor-level programs in physical and engineering sciences at the University of Cagliari (Italy).

The volumes are called 'Primers' to underline that the pedagogical aspects have been privileged over those of completeness. In particular, I selected the contents of each volume so as to keep limited its number of pages and so that the topics actually covered correspond to the *syllabus* of a typical one-semester course.

More important, however, was the choice of the style of presentation: I wanted to avoid an excessively formal treatment, preferring instead the exploration of the underlying physical features and always placing phenomenology at the centre of the discussion. More specifically, the main characteristics of this book series are:

- emphasis is always given to the physical content, rather than to formal proofs, i.e. mathematics is kept at the minimum level possible, without affecting rigour or clear thinking;
- an in-depth analysis is presented about the merits and faults of any approximation used, incorporating as well a thorough discussion of the conceptual framework supporting any adopted physical model;
- prominence is always on the underlying physical basis or principle, rather than to applications;
- when discussing the proposed experiments, the focus is given to their conceptual background, rather than to the details of the instrumental setup.

Despite the tutorial approach, I nevertheless wanted to follow the Italian academic tradition, which provides even the elementary introduction to condensed matter physics at a quantum level. I hope that my efforts have optimally combined ease of access and rigour, especially conceptual.

The intentionally non-encyclopaedic content and the tutorial character of these 'Primers' should facilitate their use even for students not specifically enrolled in a university *curriculum* in physics. I hope, in particular, that my textbooks could result accessible to students in chemistry, materials science and also of many engineering branches. In view of this, I have included a brief outline of non-relativistic quantum mechanics in the first 'Primer', a subject that does not appear in the typical engineering *curricula*. For the rest, classical mechanics, elementary thermodynamics and Maxwell theory of electromagnetism are used, to which all students of natural and engineering sciences are normally exposed.

Each 'Primer' is organised in parts, divided into chapters. This structure is tailored to facilitate the planning of a one-semester course: these volumes aim at being their main teaching tool. More specifically, each part identifies an independent teaching module, while each chapter corresponds to about two weeks of lecturing.

I cannot conclude this general introduction without thanking the many students who, over the years, have attended my courses in condensed matter physics at the University of Cagliari. Through the continuous exchange of ideas with them I have gradually understood how best to organise my teaching and the corresponding study material. As a matter of fact, the contents that I have collected in these volumes were born from this very fruitful dialogue.

Luciano Colombo
Cagliari, June 2019

Acknowledgements

I am really indebted with Ms C Mitchell—Commissioning Editor of IOP Publishing—and Mr Daniel Heatley—ebooks Editorial Assistant of IOP Publishing—for their great enthusiasm in accepting my initial proposal and for shaping it into a suitable editorial product. I also warmly acknowledge Ms Emily Tapp—Commissioning Editor of IOP Publishing—and Mr R Trevelyan—Editorial Assistant of IOP Publishing—for supporting my writing efforts of the second and the third Primers with great professionalism, always promptly replying to my queries and clarifying my doubts. Overall, their assistance has been really precious.

Author biography

Luciano Colombo

Luciano Colombo received his doctoral degree in physics from the University of Pavia (I) in 1989 and then he was a post-doc at the École polytechnique fédérale de Lausanne (CH) and at the International School for Advanced Studies (I). He became assistant professor (tenured) at the University of Milano (I) in 1990, next moving to the University of Milano-Bicocca (I) in 1996 for an equivalent position. In 1999 he was appointed associated professor at the University of Cagliari (I) where in 2002 he became full professor of theoretical condensed matter physics. Since 2015 he has been a fellow of the 'Istituto Lombardo—Accademia di Scienze e Lettere' (Milano, I). In 2021 he was appointed as Vice-Rector for Research of the University of Cagliari (I) and 'Managing Editor' of the 'Materials Physics' section of *The European Physical Journal—Plus*. He has been the principal investigator of several research projects addressed to solid-state and materials physics problems, the supervisor of more than 80 students (at bachelor, master, and PhD level), and the mentor of about 20 post-docs. He is the author, or coauthor, of more than 280 scientific articles and 9 books (this included). More about him can be found at: http://people.unica.it/lucianocolombo.

Introduction to: *Statistical Physics of Condensed Matter Systems : A primer*

This is the third and final volume in the series of three Primers I devoted to condensed matter physics. In the first two I developed the phenomenological, conceptual and formal bases for a quantum description of atoms and molecules (namely, its elementary constituents) and the crystalline state (that is: its specific form characterised by translational invariance), respectively.

A simplistic mechanistic-reductionist approach would lead one to consider the work concluded: once the fundamental principles are known, the rest is merely their systematic application to systems of increasing complexity. In fact, this way of understanding things is inefficient and badly fails to grasp a fundamental aspect of the physics of aggregates. On the one hand, as the number of elementary constituents (be they atoms or molecules) increases, a calculation that claims to explicitly consider all their degrees of freedom would soon become unmanageable, even with the aid of the most powerful high-performance computing systems. On the other hand, and even more significantly, such a brute-force approach would indeed be pointless: most of the details that dominate the atomic scale are averaged at the meso- and macro-scale, becoming irrelevant.

The conceptually important point is this: the macroscopic properties of an aggregate are just an average of single-constituent characteristics. Making a macroscopic prediction, therefore, means calculating the average value of any given physical observable which can be represented by microscopic (that is: defined at the atomic or molecular level) quantities. This is the conceptual core of the statistical description of condensed matter systems.

One might think that this challenge is already addressed by classical thermodynamics. Indeed, thermodynamics describes the thermal properties of matter (as well as their related processes) through the use of a small number of macroscopic variables, either intensive or extensive: it is, therefore, a coarse-grain theory. The statistical approach conceptually lies somewhere between the mechanistic-reductionist and thermodynamical approaches: its ambition is to describe matter at the most fundamental level (in other words: taking explicit account of its atomic-scale structure), without, however, claiming to describe in detail all the degrees of freedom associated with its elementary constituents. In practice, statistical physics predicts the macroscopic observables through appropriate ensemble averages of microscopic properties.

The statistical approach is very elegant since it provides results of paradigmatic importance through a very effective mathematical formalism; it is also effective in predicting the emergent phenomena occurring in aggregates, providing their most fundamental explanation (when dealing with atomistic properties, one can also operate at the quantum level). Because of these characteristics it cannot be missing from any tutorial introduction to condensed matter physics and for this reason I devoted the present third Primer to this topic.

This Primer is divided into two parts, respectively addressed to the statistical physics of classical and quantum systems. Each part initially contains a general chapter where the basic concepts and related mathematics are developed leading to the Boltzmann (for classical systems), Fermi–Dirac (for fermion quantum systems), and Bose–Einstein (for boson quantum systems) distribution laws, namely the three cornerstones of statistical physics. Next, the distribution laws are applied to a thorough investigation of the thermal properties of paradigmatically important systems such as the classical ideal gas, the electron gas and the phonon gas. Some specific phenomenologies are also discussed, the statistical foundations of which are treated in detail. They include, in particular, the paramagnetism, the black-body radiation (photon gas), and the Bose–Einstein condensation.

Eight appendices are added to the text, each focussed on some technical development which, at first reading, can be skipped without compromising the general understanding of the arguments developed in the main text. A bibliography is added to each chapter as a guideline for further reading. The volume contains many figures, most of which are 'conceptual', that is: they are basically intended to provide a graphical representation of the main ideas and results developed in the written part. Tables with numerical values of important physical properties are included as well, in the attempt to provide the reader with information useful to 'quantify' the physical results presented. Finally, a list of all the mathematical symbols used in the volume is given at the beginning as an orientation guide while reading.

I started writing the first of three Primers during the Fall 2018, encouraged by my wife, no longer with us now. As I close this experience, my thoughts and deepest gratitude go out to her.

Luciano Colombo
Cagliari, July 2022

Acknowledgements

I am really indebted with many friends who helped me by critically reading (in part or totally) the pre-editorial version of the book (in particular, Dr A Cappai, University of Cagliari, Italy). The care these friends have put in checking my original manuscript has corrected several unclear passages and many misprints, thus greatly improving my presentation. If there are still errors or omissions they should be attributed solely to me.

Symbols

λ	vibrational quantum number for a quantum harmonic oscillator
μ_B	Bohr magneton
μ_c	chemical potential
τ_e	scattering relaxation time for electrons
τ_{sq}	scattering relaxation time for phonons
χ_C	Curie paramagnetic susceptibility
Ω	the number of different ways a partition can be realised
A	atomic mass number
$\mathbf{B}$	magnetic field
C_P	molar constant-pressure heat capacity
C_V	molar constant-volume heat capacity
e	electron charge
$\mathbf{E}$	electric field
E_F	Fermi energy
$f(E, T)$	Fermi–Dirac occupation number
$\mathcal{F}$	Helmholtz free energy
g	Lande' g-factor
$G(E)$	density of energy levels
$G(\nu)$	spectral density of the electromagnetic radiation
$\mathcal{G}$	Gibbs free energy
h	Planck constant
$\hbar$	normalised Planck constant $\hbar = h/2\pi$
$\mathcal{H}$	enthalpy
j	quantum number associated to the atomic total (orbital+spin) angular momentum
$\mathbf{J}$	total (orbital + spin) atomic angular momentum
k_B	Boltzmann constant
l	quantum number associated to the atomic orbital angular momentum
m_e	electron mass
m_j	total magnetic quantum number
$\mathcal{M}$	magnetisation (magnetic moment per unit volume)
n	number of moles
$n(E, T)$	Bose–Einstein occupation number
$\mathcal{N}_A$	Avogadro number
$\langle O \rangle$	ensemble average of the observable O
P	pressure
P_{BB}	total black-body electromagnetic power emitted per unit area
$\mathcal{P}$	polarisation (electric moment per unit volume)
$\mathcal{Q}$	heat
r	rotational quantum number for a diatomic molecule
R	universal gas constant
s	quantum number associated to the spin angular momentum
S	Boltzmann entropy
S_{Gibbs}	Gibbs entropy
T	temperature
T_{BE}	transition temperature for the Bose–Einstein condensation
T_F	Fermi temperature (electron gas)

T_{room}	room temperature (such that $k_{\text{B}}T_{\text{room}} = 0.025$ eV)
T_{rot}	rotational temperature (diatomic molecule)
T_{vib}	vibrational temperature (diatomic molecule)
$u_{\text{BB}}(\nu)$	black-body spectral energy density
$\mathcal{U}$	total internal energy
$\mathcal{U}_{\text{rot}}$	rotational internal energy of a gas
$\mathcal{U}_{\text{tr}}$	translational energy of a gas
$\mathcal{U}_{\text{vib}}$	vibrational energy of a gas or crystalline solid
v	particle velocity (ideal gas)
v_{mp}	particle most probable velocity (ideal gas)
$\langle v \rangle$	particle average velocity (ideal gas)
$\langle v^2 \rangle$	particle mean square velocity (ideal gas)
V	volume
Z_{v}	number of valence electrons per atom
$\mathcal{Z}$	partition function
$\mathcal{Z}_{\text{GC}}$	grand canonical partition function
$\mathcal{Z}_{\text{rot}}$	rotational partition function of a gas
$\mathcal{Z}_{\text{vib}}$	vibrational partition function of a gas
$\overline{\mathcal{Z}}$	grand partition function
W	mechanical work

Part I

Classical statistical physics

IOP Publishing

Statistical Physics of Condensed Matter Systems
A primer
Luciano Colombo

Chapter 1

The statistical description of a classical system

Syllabus—*We introduce the basic concepts underlying the statistical description of a system made by a large number of classical (spinless) particles. At first, two fundamental challenges are taken up, namely how to find the most probable distribution law for its equilibrium state and how to estimate the corresponding macroscopic properties. Next, the concept of entropy is thoroughly presented and used to elaborate a robust microscopic foundation of classical thermodynamics. Eventually, the thermal physics of the monoatomic ideal gas is discussed, in particular deriving its equation of state and the Maxwell distribution laws for energy and molecular velocities. The statistical definition of irreversible process concludes this chapter.*

1.1 Basic concepts

Our modern physical picture of condensed matter defines ordinary systems as an assembly of interacting atoms or molecules [1, 2]. This holds for any aggregate, either solid, liquid or gaseous. According to this *atomistic picture*, any observed macroscopic property can be intended as the result of organised actions occurring among a really huge[1] number of elementary constituents. However, any phenomena emerging at the macroscale can hardly be described by considering each single constituent and its interplay with all the remaining ones, simply because there are too many degrees of freedom to be managed. Even more important: pretending to follow each atom or molecule would be a rather clumsy approach, since most of the atomic-scale details are in fact averaged out and, therefore, their overall inclusion in a physical model would be a pedantic solution adding no meaningful knowledge.

---

[1] In normal temperature and pressure conditions, one cm^3 of a gas typically contains as many as $\mathcal{O}(10^{19})$ constituents.

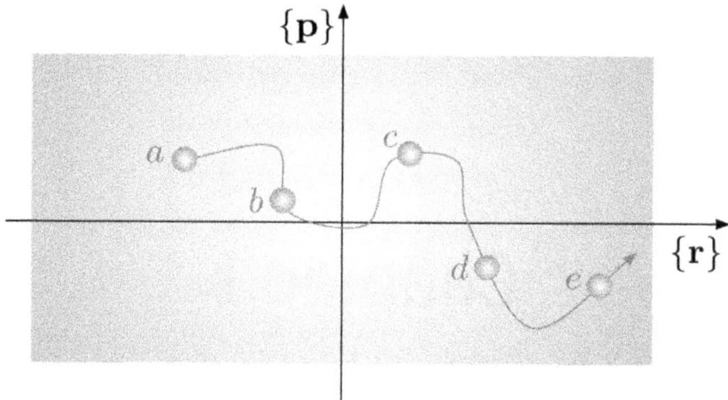

Figure 1.1. Pictorial representation of the phase space of a classical system. Different microstates $a, b, \ldots, e$ are shown. The sequence $a \to b \to \cdots \to e$ defines the trajectory in the phase space followed by the system during its time evolution. Axes define the full set of momenta $\{\mathbf{p}\}$ and positions $\{\mathbf{r}\}$ of the particles forming the system.

A rather different approach will be developed in this Primer, where processes involving very many elementary constituents (whose mutual interactions are known) will be described without considering the individual behaviour of each of them. We name this approach the *statistical description of a condensed matter system*. Basically, we aim at understanding the collective or macroscopic properties of an aggregate, without describing the motion of its constituents in full detail. The key point of this statistical description is that, while the atomistic picture is still adopted in describing the nature of the given aggregate, the single-particle mechanical approach will be replaced by a statistical one, providing predictions for the average (macroscopic) value of physical quantities.

In order to explain the basic concepts underlying statistical physics more rigorously, let us consider a classical system containing a *large number N* of particles. In this framework by 'classical system' we mean any aggregate of *spinless particles*; also, we will make no use of its quantum mechanical wavefunction in addressing the description of its physical state. If we assign an exact value to the space coordinates and momenta of each particle we define a specific *microstate* of the system. Such a microstate is associated with a single point in the $6N$-dimensional *phase space* of the system, whose axes represent all the coordinates and momenta, as shown in figure 1.1. If the system state undergoes a time evolution, its representative point follows a trajectory in the phase space. For the system to follow a specific trajectory, a small number of thermodynamical (that is macroscopic) variables must be assigned; for instance, we might fix the total internal energy $\mathcal{U}$, the volume V, and the total number of particles N of the system[2]. This reduced set of parameters defines the *macrostate* of the system. Formally this is equivalent to saying that we are

---

[2] If we are dealing with a multi-component system, that is, a system containing two or more chemical species, we should define the number of particles for each chemical species. This is just an annoying detail, which does not affect our reasoning and, therefore, will be neglected.

forcing the trajectory to lie on a specific surface of the phase space, defined by the assigned thermodynamical parameters.

If we wait a long enough time, it is reasonable to assume that the system will fully explore the surface defining a macrostate, that is, *it will visit all the microstates consistent with the imposed macroscopic constraints*. The actual value of any physical observable is simply the average among all the values it assumed in each visited microstate. The averaging operation is referred to as an *ensemble average*, where the word 'ensemble' simply identifies the assembly of all microstates. This situation is summarised by stating that *an ensemble average is the same as a time average*. Systems obeying this condition are said to be *ergodic*. While this subtle concept is treated more in detail elsewhere [3–5], here we remark a key notion of the statistical approach, namely: under the ergodic assumption, the observed value of a property corresponds equally well to a time average performed over a long enough time or to an ensemble average performed over a rich enough assembly of microstates[3]. In other words, we could select a specific microstate, follow its dynamical evolution on the surface defined by the assigned macroscopic parameters, and average over the values it assumed in the visited microstates; alternatively, we could randomly select a large enough number of microstates on the same surface and take the average over them. The result will be just the same[4].

In order to proceed, we need to establish a link between a macrostate and its corresponding microstates. By keeping the situation as clean as possible, let us assume that the N particles forming the system are identical and structureless. Let us further assume that they can occupy states with energy E_i with $i = 1, 2, 3,\dots$ (with $E_1 < E_2 < E_3 < \dots$). Then, it is straightforward to write

$$N = \sum_i n_i \quad \text{and} \quad \mathcal{U} = \sum_i n_i E_i \tag{1.1}$$

where n_i is the number of particles with energy E_i; this number is unrestricted since we are treating an aggregate of classical (spinless) particles. We need to remark that, strictly speaking, the above definition of internal energy is only valid for a system of non-interacting particles, since only in this case can we unambiguously attach to each particle its energy; however, to a good approximation, it could be extended to a system of weakly interacting constituents (like e.g. a dilute gas at high temperature). In order to manage the case where interactions are not negligible, we could adopt a *single-particle approximation* [2]: the many-body problem is reduced to N single-particle problems, where each ith particle undergoes a self-consistent potential V_i^{scp} only depending on its

---

[3] It is understood that both averages are taken under the same thermodynamical constraints.

[4] This conclusion has very important practical consequences. By implementing a computational approach to statistical physics, we could equally well perform ensemble averages by following the deterministic trajectory (we are developing a classical theory!) of the representative point of the system in the phase space or by randomly selecting microstates lying on the same surface. The first approach is named 'molecular dynamics', the second one 'Monte Carlo method' [6]: they are the kernel of contemporary computational statistical physics.

coordinates. Such potential V_i^{scp} is local and effectively describes the interaction with all the remaining $(N-1)$ particles in the system. Under this approximation $E_i = E_i^{\text{kin}} + V_i^{\text{scp}}$ where both the kinetic and the potential contribution to the single-particle energy appear. For strongly interacting systems where the single-particle approximation fails, other techniques must be developed, explicitly including interactions. This situation falls beyond the scope of this Primer.

If the system is isolated, its energy $\mathcal{U}$ is conserved. However, collisions[5] and interactions affect the *distribution of the particles among the energy states* which is referred to as *partition*; this implies that the occupation numbers n_i may change in time and the partition (or, equivalently, the microstate) may accordingly change. A natural question arises at this point: *which is the most probable partition, once the macrostate[6] of the system has been assigned?* It is of fundamental importance to answer this question, since *a system in its most probable partition is said to be at equilibrium.* This statement can be easily justified since it is quite obvious to assume that the free overall evolution of an isolated system is always from a partition with lower probability to a partition with higher probability.

When an equilibrium condition is reached, there is no change of macrostate unless a disturbing external action is applied to the system. On the other hand, statistical fluctuations may occasionally affect the partition (or, equivalently, the microstate), but they are ultimately unable to lead the system far from equilibrium: in other words, the n_i can fluctuate around their equilibrium values, without sizeable macroscopic effects. Once the fundamental question of statistical physics is solved[7] the next challenge is to develop theoretical methods for *predicting, by means of an ensemble average, the equilibrium values of the physical properties we are interested in.*

In concluding this introduction to the statistical approach, it is important to remark that the use of average-based arguments for predicting system properties does not mean that the physical picture derived from it is vague, or approximate or even just qualitative. On the contrary, we will develop a very quantitative and robust description of physical systems containing a large number of constituents. Rather, we should more correctly state that, whenever an observable is specified by a reduced number of macroscopic parameters, it will be described statistically, through averaging procedures. This is the reason why statistical physics is the basic theory underlying macroscopic thermodynamics. The first theoretical formulation of the statistical approach was developed by L Boltzmann, J C Maxwell, and J W Gibbs within classical physics in the years between the XIX and the XX century.

---

[5] Occurring even within a non-interacting system.

[6] That is, for the case we discussed so far, once its volume, number of particles, and energy have been assigned.

[7] In other words: once, given an isolated system with known composition, its most probable partition has been found.

1.2 Equilibrium distribution probability

In order to develop the quantitative calculation of the equilibrium partition for a classical system, we will restrict ourselves to considering a set of identical particles which, at least for the moment, will be considered also *distinguishable*[8].

Let us then consider an isolated system of N such *identical* and *distinguishable* particles and let us further assume that all their allowed states with energy E_i have the *same probability of being occupied*. It is important to remark that both this assumption and the distinguishability requirement will be soon removed, in order to elaborate a more general picture; at this stage they must be intended just as pedagogical tools. Any given partition can be realised in different ways, each corresponding to a specific distribution of the particles among the states. It is quite reasonable to consider that *the probability of each partition is proportional to the number of different ways it can be realised*. The statistical theory we are going to develop is basically funded on this consideration. It is therefore mandatory to correctly count the number of distinguishable different ways we can place $n_i \leqslant N$ particles on the state with energy E_i, once the remaining particles have been already placed.

If we start from the lowest level E_1 such a number is $N!/n_1!(N - n_1)!$, corresponding to the number of permutations of N objects taken n_1 at a time. For the second state with energy E_2 the number of available particles to accomodate is $(N - n_1)$ and, therefore, it is easy to understand that $(N - n_1)!/n_2!(N - n_1 - n_2)!$ is the number of distinguishable different ways we can place n_2 particles on it. By iterating this procedure for all the remaining states, we eventually obtain *the number of different ways a partition can be realised with n_1 particles on the state E_1, n_2 on the state E_2, n_3 on the state E_3, and so on*

$$\frac{N!}{n_1!n_2!n_3!\ldots} \tag{1.2}$$

while, as stated above, the corresponding probability P for this partition is proportional to it.

We proceed by refining and generalising our reasoning and, first of all, we admit that *the allowed states have now different occupation probabilities which we will indicate by p_i*. A direct way to understand this is to admit the energy E_i is degenerate: states with higher degeneracy are more likely to be occupied. The true existence of different occupation probabilities implies that the number of different ways we can realise the above partition must be corrected as

$$\frac{N!p_i^{n_1}p_2^{n_2}p_3^{n_3}\ldots}{n_1!n_2!n_3!\ldots} \tag{1.3}$$

---

[8] More specifically, two particles are identical if they have the same structure and physical properties (two H atoms or two NaCl molecules in the same quantum state are so), while they can be distinguished for instance by looking at their position (at least according to classical physics).

since the probability of finding n_i particles on the state E_i is precisely $p_i^{n_i}$. It is very important to remark that *we are not applying any restriction on such intrinsic (or state) probabilities p_i*. In other words, we are not assuming any exclusion principle related to the character of the total wavefunction. This is tantamount to stating that *we are developing a purely classical statistical theory*.

Next, we remove the distinguishability assumption (in other words, particle labelling is any longer relevant): this implies that the $N!$ permutations among particle pairs that occupy the different states actually provide the very same partition. Therefore, *the number Ω of different ways the partition can be realised* is eventually corrected in

$$\Omega = \frac{1}{N!} \frac{N! p_i^{n_1} p_2^{n_2} p_3^{n_3} \cdots}{n_1! n_2! n_3! \cdots} = \frac{p_i^{n_1} p_2^{n_2} p_3^{n_3} \cdots}{n_1! n_2! n_3! \cdots} = \prod_i \frac{p_i^{n_i}}{n_i!} \tag{1.4}$$

which corresponds to a probability

$$P = \xi \prod_i \frac{p_i^{n_i}}{n_i!} \tag{1.5}$$

where ξ is a convenient normalisation factor linking the number of different ways the partition can be realised and its corresponding probability $P \in [0.1]$. This factor does not play any role in the theory, as it will be clear very soon.

According to the definition provided in section 1.1, the equilibrium state corresponds to the most probable partition and, therefore, it can be calculated by maximising the quantity provided in equation (1.5). This procedure, however, must fulfil two physical constraints provided in equations (1.1), namely (i) the conservation of energy and (ii) the conservation of the number of particles. In practice, the calculation proceeds by considering at first the natural logarithm[9] of the partition probability P

$$\ln P = \ln \xi + \ln \left(\prod_i \frac{p_i^{n_i}}{n_i!} \right) = \ln \xi + \sum_i \ln \frac{p_i^{n_i}}{n_i!} = \ln \xi + \sum_i \left(n_i \ln p_i - \ln n_i! \right) \tag{1.6}$$

which is easily calculated by means of the Stirling formula $\ln n! \simeq n \ln n - n$ valid for large enough n numbers (see appendix A)

$$\ln P = \ln \xi + \sum_i \left(-n_i \ln \frac{n_i}{p_i} + n_i \right) = \ln \xi + N - \sum_i n_i \ln \frac{n_i}{p_i} \tag{1.7}$$

Next, in order to find its maximum we must differentiate with respect to small changes dn_1, dn_2, dn_3,... of the occupation numbers

$$-d \ln P = \sum_i dn_i \ln \frac{n_i}{p_i} = 0 \tag{1.8}$$

---

[9] Since the logarithm is a monotonic function of its variable, the maximum of $\ln P$ corresponds to the maximum of P.

where we have taken into consideration that both ξ and N are constant since, respectively, the first one is just a numerical factor and the system is isolated. This latter condition further imposes that upon the same small changes we must have

$$\sum_i dn_i = 0 = \sum_i E_i\, dn_i \tag{1.9}$$

so that the maximum condition for the quantity appearing in equation (1.8) must be compensated according to the Lagrange method (see appendix A)

$$\sum_i \left(\ln \frac{n_i}{p_i} + \alpha + \beta E_i \right) dn_i = 0 \tag{1.10}$$

where α and β are the Lagrange multipliers. The maximum probability is then found by imposing

$$\ln \frac{n_i}{p_i} + \alpha + \beta E_i = 0 \tag{1.11}$$

which leads to

$$n_i = p_i \exp(-\alpha - \beta E_i) \tag{1.12}$$

the fundamental result paving the way to the statistical mechanics of classical identical particles.

1.3 The Boltzmann distribution law

In order to proceed further we need to determine the actual value of the Lagrange multipliers α and β appearing in equation (1.12). To this aim, let us calculate explicitly the total number of particles as

$$N = \sum_i n_i = \sum_i p_i \exp(-\alpha - \beta E_i) = \exp(-\alpha) \sum_i p_i \exp(-\beta E_i) \tag{1.13}$$

and introduce the *partition function* Z

$$Z = \sum_i p_i \exp(-\beta E_i) \tag{1.14}$$

which represents a key quantity of statistical physics. By combining equations (1.13) and (1.14) we obtain $\exp(-\alpha) = N/Z$ and eventually write

$$n_i = \frac{N}{Z} p_i \exp(-\beta E_i) \tag{1.15}$$

which is known as *the Boltzmann distribution law* describing the equilibrium statistics of an assembly of classical identical particles distributed on energy levels E_i with intrinsic probabilities p_i.

1.4 A statistical definition of temperature

For equation (1.15) to be a useful result for applications we still need to provide our theory with an explicit and computable expression of the β multiplier. To this aim we observe that for dimensional consistency β must be measured in reciprocal energy units; therefore, we guess that working on the total energy of the system could be the right way to proceed.

By inserting equation (1.15) into equation (1.1) we get

$$\mathcal{U} = \frac{N}{\mathcal{Z}}\sum_i p_i E_i \exp(-\beta E_i) = -\frac{N}{\mathcal{Z}}\frac{d}{d\beta}\left[\sum_i p_i \exp(-\beta E_i)\right] \tag{1.16}$$

which allows to express *the energy of a system of particles in terms of its partition function* as

$$\mathcal{U} = -N\frac{d}{d\beta}(\ln \mathcal{Z}) \tag{1.17}$$

and the *average energy per particle* as

$$\langle E \rangle = -\frac{d}{d\beta}(\ln \mathcal{Z}) \tag{1.18}$$

where $\langle E \rangle = \mathcal{U}/N$.

While equations (1.17) and (1.18) clearly confirm our guess that *the parameter β is linked to the system internal energy*, for historical reasons it has been preferred to introduce a new physical quantity T, referred to as *absolute temperature*, defined as

$$k_{\mathrm{B}}T = \frac{1}{\beta} \tag{1.19}$$

where k_{B} is a suitable constant, discussed below, so that the product $k_{\mathrm{B}}T$ is expressed in energy units. Equation (1.19) represents the *statistical definition of temperature*: it is a positive-definite quantity (the lowest absolute temperature value is therefore zero), whose relation with the phenomenological temperature defined in classical thermodynamics will soon be established.

It is important to remark that the absolute temperature introduced through the above statistical arguments is only defined at equilibrium: the fundamental reason is that the definition of T is based on the β parameter which, in turn, has been elaborated when looking for the most probable distribution of the system particles among the allowed energy state or, more concisely, when looking for the equilibrium condition of the system[10].

---

[10] In non-equilibrium situations is still possible to define a 'local temperature' for each portion of the system which is still large enough to validate a statistical approach, but comparatively much smaller than the full system size. This allows, for example, defining how the temperature varies within a material system in stationary condition, along the direction of an applied thermal gradient: indeed a non-equilibrium situation! While this comment is only qualitative, non-equilibrium thermodynamics makes this definition of temperature quite robust through the concept of 'local equilibrium' [7].

The physical meaning of the statistical temperature defined in equation (1.19) is fully understood by considering a system formed by two interacting subsystems containing $N^{(a)}$ and $N^{(b)}$ particles, respectively, so that $N = N^{(a)} + N^{(b)}$. Different energy states are allowed for the particles in the two subsystems, hereafter labelled as $E_i^{(a)}$ and $E_j^{(b)}$. We will consider a simplified, but meaningful, situation in which (i) the system as a whole is isolated, (ii) the two subsystems exchange energy through particle–particle interactions and collisions, and (iii) no phenomena could occur affecting the particle numbers. In short: the system total energy is conserved, as well as each individual particle number $N^{(a)}$, and $N^{(b)}$; on the other hand, the energy of each subsystem can vary. This situation is described by the following three constraints

$$N^{(a)} = \sum_i n_i^{(a)} \quad N^{(b)} = \sum_j n_j^{(b)} \quad \mathcal{U} = \sum_i n_i^{(a)} E_i^{(a)} + \sum_j n_j^{(b)} E_j^{(b)} \tag{1.20}$$

while, by extending the arguments developed in section 1.2, the probability P to observe a given partition with $n_i^{(a)}$ particles of the first subsystem on the energy levels $E_i^{(a)}$ and $n_j^{(b)}$ particles of the second subsystem on the energy levels $E_j^{(b)}$ is given by

$$P = \xi \left[\prod_i \frac{[p_i^{(a)}]^{n_i^{(a)}}}{n_i^{(a)}!} \right] \left[\prod_j \frac{[p_j^{(b)}]^{n_j^{(b)}}}{n_j^{(b)}!} \right] \tag{1.21}$$

where $p_i^{(a)}$ and $p_j^{(b)}$ are the occupation probabilities of the various energy states in the two subsystems, respectively. This equation is the straightforward extension of equation (1.5). In order to find the equilibrium distribution for the system we will follow the same procedure developed in section 1.2.

By differentiating the natural logarithm of the probability given in equation (1.21) with respect to small changes $dn_i^{(a)}$ and $dn_j^{(b)}$ in the occupation numbers within the two subsystems, we get

$$-d(\ln P) = \sum_i \ln \frac{n_i^{(a)}}{p_i^{(a)}} \, dn_i^{(a)} + \sum_j \ln \frac{n_j^{(a)}}{p_j^{(b)}} \, dn_j^{(b)} = 0 \tag{1.22}$$

where we set once again to zero such a variation since we are looking for the maximum probability distribution. The conservation conditions are now dictated as

$$\sum_i dn_i^{(a)} = 0 \quad \sum_j dn_j^{(b)} = 0 \quad \sum_i E_i^{(a)} dn_i^{(a)} + \sum_j E_j^{(b)} dn_j^{(b)} = 0 \tag{1.23}$$

leading to

$$\sum_i \left(\ln \frac{n_i^{(a)}}{p_i^{(a)}} + \alpha^{(a)} + \beta E_i^{(a)} \right) dn_i^{(a)} + \sum_j \left(\ln \frac{n_j^{(b)}}{p_j^{(b)}} + \alpha^{(b)} + \beta E_j^{(b)} \right) dn_j^{(b)} = 0 \tag{1.24}$$

once the three multipliers $\alpha^{(a)}$, $\alpha^{(b)}$, and β have been introduced. The above equation is satisfied provided that

$$\ln \frac{n_i^{(a)}}{p_i^{(a)}} + \alpha^{(a)} + \beta E_i^{(a)} = 0 = \ln \frac{n_j^{(b)}}{p_j^{(b)}} + \alpha^{(b)} + \beta E_j^{(b)} \tag{1.25}$$

or equivalently

$$n_i^{(a)} = \frac{N^{(a)}}{\mathcal{Z}^{(a)}} p_i^{(a)} \exp[-E_i^{(a)}\beta E_i^{(a)}] \quad \text{and} \quad n_j^{(b)} = \frac{N^{(b)}}{\mathcal{Z}^{(b)}} p_j^{(b)} \exp\left[-E_j^{(b)}\beta E_j^{(b)}\right] \tag{1.26}$$

where $\mathcal{Z}^{(a)}$ and $\mathcal{Z}^{(b)}$ are the respective partition functions of the two subsystems

$$\mathcal{Z}^{(a)} = \sum_i p_i^{(a)} \exp[-\beta E_i^{(a)}] \quad \text{and} \quad \mathcal{Z}^{(b)} = \sum_j p_j^{(b)} \exp\left[-\beta E_j^{(b)}\right] \tag{1.27}$$

which, interestingly enough, result in *having the same value of the parameter β*. In other words, thanks to this result and to the definition provided in equation (1.19) we conclude that *in equilibrium the two subsystems have the very same absolute temperature*. This is perfectly consistent with the *zeroth law of classical thermodynamics* [3, 8–10] and, therefore, makes the statistical definition of temperature fully compatible with its phenomenological counterpart. According to the present argument, it is understood that in equilibrium conditions the two subsystems may exchange energy at the microscopic level (occupation numbers in both subsystems do vary around their equilibrium values because of mutual interactions); however, such an exchange occurs in both directions so that statistically their energy remains constant. On the other hand, if the two subsystems are initially at different temperature, interactions cause energy exchanges until the same-temperature equilibrium condition is eventually reached.

The zeroth law has an important practical consequence. Its equivalent verification at the statistical and empirical level implies that *the statistical absolute temperature of a system is precisely the experimentally detected one*, since this latter is determined by placing in contact the system with a thermometer and waiting until equilibrium is reached. We recall that a thermometer is a device characterised by a thermometric parameter, that is, by a physical temperature-dependent property like, for instance, its volume or electrical resistance [9]. Different values of such a property are collected at corresponding conventional situations (for instance, at freezing and boiling conditions of water at normal pressure) and a suitable temperature scale is accordingly defined. Provided that absolute temperature is measured in degrees Kelvin, the conventional value of the constant k_{B} introduced in equation (1.19) is

$$k_{\mathrm{B}} = 1.3805 \times 10^{-5} \text{ eV K}^{-1} \tag{1.28}$$

and it is commonly referred to as the *Boltzmann constant*.

1.5 Calculating ensemble averages

Let us consider a system containing N particles and an observable O which can be expressed as a function of the particle energies E_i. Then, for any given partition its *average value* $\langle O \rangle$ is by definition calculated as

$$\langle O \rangle = \frac{1}{N} \sum_i n_i O(E_i) \tag{1.29}$$

which, by using the equilibrium distribution probability provided in equation (1.15) and the temperature definition provided in equation (1.19), is transformed into

$$\langle O \rangle = \frac{1}{\mathcal{Z}} \sum_i p_i \, O(E_i) \, \exp(-E_i/k_\mathrm{B}T) \tag{1.30}$$

corresponding to *the operational definition of ensemble average.*

A very special observable is of course the internal energy $\mathcal{U}$ defined in equation (1.17). Since equation (1.19) holds, we have

$$d\beta = -\frac{dT}{k_\mathrm{B}T^2} \tag{1.31}$$

and we obtain the *link between the partition function and the internal energy* as

$$\mathcal{U} = k_\mathrm{B}NT^2 \frac{d}{dT} \ln \mathcal{Z} \tag{1.32}$$

while the average energy per particle is

$$\langle E \rangle = k_\mathrm{B}T^2 \frac{d}{dT} \ln \mathcal{Z} = \frac{1}{\mathcal{Z}} \sum_i p_i E_i \exp(-E_i/K_\mathrm{B}T) \tag{1.33}$$

where we made use of equation (1.18). This result, fully consistent with the above definition of ensemble average, is of great conceptual relevance since *it establishes a direct link between the system temperature and the average energy per particle.* Using a more phenomenological language, by virtue of equation (1.33) we can say that the sensorial experience of 'hot' and 'cold' is nothing other than an effective (macroscopic) measure of the energy content of the system we entered in contact with.

We conclude this section by remarking that any explicit calculation of ensemble averages does require the knowledge of the partition function, which depends on the microscopic structure of the system as well as on the thermodynamical conditions the system is subjected to. Basically, the statistical description of a physical system is a two-step procedure: the partition function must be at first calculated; next, any observable can be straightforwardly evaluated according to equation (1.30). More specifically, different partition functions must be in principle considered for an isolated system or a system in contact with a thermal bath or a system which can vary its matter content. Averages calculated in the first case (where the system is characterised by a fixed energy and number of particles) correspond to the *micro-canonical ensemble*, while in the second case (where the system can exchange energy, while its particles are still fixed in number) they define the *canonical ensemble*. In both the microcanonical and canonical case, the partition function $\mathcal{Z}$ is given by equation (1.14). In contrast, if the system can exchange energy and mass (that is, the number of particles is no longer fixed) the corresponding statistics is performed in the *grand canonical ensemble* for which a different form of the partition function must be elaborated. This topic falls beyond the scope of this Primer[11], as well as the details of the ensemble theory which can be found elsewhere [3, 10, 11]. The key point to retain for the following is that *the fundamental task of statistical physics is to evaluate the partition function of any specific system in assigned thermodynamical conditions*. Only after this mission has been accomplished, predictions about the (average) system properties can be elaborated.

1.6 Entropy

1.6.1 Thermodynamical entropy

In thermodynamics a function named *entropy S* is defined for any system in whatever equilibrium as well as non-equilibrium state, such that its variation when the system evolves depends only on the initial and final states occupied by the system and not on the specific process followed. Being a state function, S dictates the thermal physics of the system: any thermodynamical parameter can be calculated from it [8, 9, 11]. For instance, temperature T, pressure P and chemical potential μ_c, namely the three mostly relevant *intensive quantitites*, are directly calculated once the entropy of the system is known

$$\frac{1}{T} = \frac{\partial S}{\partial \mathcal{U}}\bigg|_{V,N} \qquad \frac{P}{T} = \frac{\partial S}{\partial V}\bigg|_{\mathcal{U},N} \qquad -\frac{\mu_c}{T} = \frac{\partial S}{\partial N}\bigg|_{\mathcal{U},V} \tag{1.34}$$

where for sake of simplicity we have consider a mono-component system[12] whose entropy depends on just three parameters $S = S(\mathcal{U}, V, N)$. For an infinitesimal process we can therefore write

$$dS = \frac{1}{T}d\mathcal{U} + \frac{P}{T}dV - \frac{\mu_c}{T}dN \tag{1.35}$$

from which we immediately obtain the *first law of thermodynamics*: the variation of internal energy of the system

$$d\mathcal{U} = TdS - PdV + \mu_c dN = d\mathcal{Q} - dW + \mu_c dN \tag{1.36}$$

---

[11] The only exception is treated in section 3.4.

[12] The extension to multi-component systems is straightforward, but since it does not add any new physical feature it will not be considered here.

is given by the balance between the amount of heat $dQ = TdS$ exchanged with the environment[13], the mechanical work $dW = P\,dV$ caused by volume variations, and the chemical work $\mu_c\,dN$ due to the variation in the particle number of the system.

1.6.2 Boltzmann entropy

Entropy being such a fundamental object of thermodynamics, it deserves a *statistical definition*, which was in fact guessed by Boltzmann in the form

$$S = k_B \ln \Omega \tag{1.37}$$

where Ω is the number of microstates compatible with the selected macrostate of the system (sometimes Ω is also referred to as the 'multiplicity function'). It must be clear that *the definition provided in equation* (1.37) *is just a postulate* [3, 5], whose validity can only be proved heuristically, as thoroughly discussed below. Nevertheless, it is supported by robust and meaningful arguments. First of all, we observe that when a system evolves towards equilibrium Ω is maximised (see discussion in section 1.1): therefore, according to equation (1.37) *S is maximum for the equilibrium state*, consistently with thermodynamics [8, 9]. Next, we recall that statistical laws are multiplicative and, therefore, if a system is composed by two non-interacting subsystems, respectively, characterised by Ω_1 and Ω_2 number of microstates, it has $\Omega_1\Omega_2$ partitions. This implies that the entropy of the total system is $S = k_B \ln(\Omega_1\Omega_2) = k_B \ln \Omega_1 + k_B \ln \Omega_2 = S_1 + S_2$. In other words, equation (1.37) states that *S is an extensive quantity*, one again as dictated by thermodynamics [8, 9]. A more formal argument leading to the same definition as in equation (1.37) is outlined in section 1.6.3, where the concept of entropy is developed as *a measure of our ignorance on the actual microstate occupied by the system* once its macrostate has been assigned.

Let us now consider an isolated system out of equilibrium and follow its natural evolution: *we always observe a positive entropy variation* (simply because the free evolution drives the system to the macrostate with maximum number of realisations). On the other hand, if the isolated system was already at equilibrium, the only observed spontaneous processes are those characterised by a zero entropy variation. These arguments are tantamount to stating that *in isolated systems we always have* $dS \geqslant 0$, corresponding to the second law of thermodynamics[14]; $dS = 0$ processes are called *reversible*[15], while $dS > 0$ ones are said to be *irreversible*. The statistical nature

---

[13] We remark that $dQ = T\,dS$ holds only for quasi-static processes, that is for processes occurring through a sequence of equilibrium states, where the absolute temperature T can be defined. In practice, they correspond to evolutions proceeding by infinitesimal steps occurring over a time scale much longer than the time needed for the system to reach a new equilibrium state (also referred to as relaxation time). The relation $dQ = T\,dS$ was introduced for the first time by Clausius, long before the microscopic approach to thermal physics was elaborated; it represents the macroscopic operational definition for the entropy variation.

[14] If the system is not isolated, it is indeed possible to observe a negative entropy variation; however, $dS \geqslant 0$ for the larger complex formed by the system and its embedding environment.

[15] The notion of reversible process and of quasi-static process are here considered equivalent.

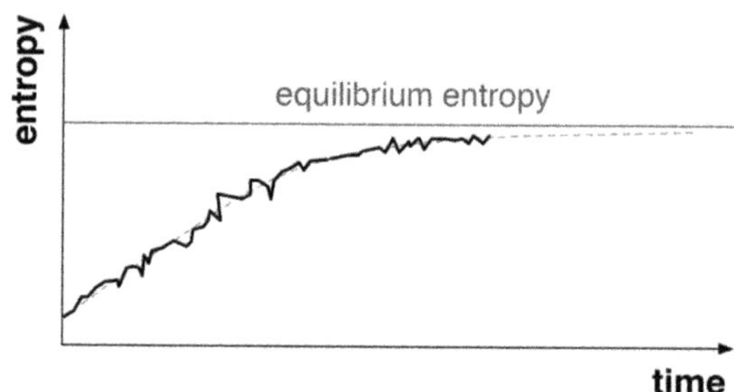

Figure 1.2. Entropy variation for an isolated system evolving towards equilibrium. Dashed line: a guide to the eye, corresponding to a macroscopic thermodynamical prediction. Noisy full line: actual entropy variation according to equation (1.37).

of the definition reported in equation (1.37) allows us to admit that, because of fluctuations, the entropy of an isolated system may occasionally decrease; however, the likelihood of such negative variations is inversely proportional to their amplitude. In figure 1.2 a pictorial representation of the entropy variation towards equilibrium for an isolated system is sketched.

Let us now proceed with the explicit calculation of the entropy for a classical system obeying the Boltzmann statistics. By inserting equation (1.4) into equation (1.37) and using the Stirling formula (see appendix A) we get

$$S = k_B \left(\sum_i n_i \ln p_i - \sum_i n_i \ln n_i + \sum_i n_i \right) = -k_B \sum_i n_i \ln \frac{n_i}{p_i} + k_B N \qquad (1.38)$$

where we used equation (1.1) for the total number of particles. The logarithm appearing in the sum on the right hand side of this equation is easily calculated by using equation (1.15)

$$S = k_B \sum_i n_i \frac{E_i}{k_B T} + k_B \sum_i n_i \ln \frac{\mathcal{Z}}{N} + k_B N = \frac{1}{T} \sum_i n_i E_i + k_B \ln \frac{\mathcal{Z}}{N} \sum_i n_i + k_B N \qquad (1.39)$$

where it is easy to recognise the total internal energy of the system and the total number of particles in the two sums appearing, respectively, in the first and second term of the right hand side of this equation. By means of equations (1.1) we therefore obtain

$$S = \frac{U}{T} + k_B N \ln \frac{\mathcal{Z}}{N} + k_B N \qquad (1.40)$$

which eventually provides

$$S = \frac{U}{T} + k_B \ln \frac{\mathcal{Z}^N}{N!} \qquad (1.41)$$

as the *entropy for a system of identical and indistinguishable particles following the Boltzmann statistics.*

The Boltzmann formulation for the entropy can justify the Clausius definition $dQ = TdS$ linking the amount of exchanged heat to the entropy variation observed during a reversible process. This result is so relevant that it deserves a proof, which is interesting either as a showcase application of the statistical formalism and for better establishing its link to thermodynamics. If we consider a reversible transformation of an isolated mono-component system obeying the Boltzmann statistics and use equation (1.40), we can write

$$dS = \frac{1}{T}d\mathcal{U} - \frac{\mathcal{U}}{T^2}dT + \frac{k_{\mathrm{B}}N}{\mathcal{Z}}d\mathcal{Z} \tag{1.42}$$

where, since the system is isolated, we set $dN = 0$. We now focus on the last term appearing on the right hand side of this equation and we calculate at first

$$d\mathcal{Z} = -\sum_i \frac{p_i}{k_{\mathrm{B}}T}\exp(-E_i/k_{\mathrm{B}}T)dE_i + \sum_i \frac{p_i E_i}{k_{\mathrm{B}}T^2}\exp(-E_i/k_{\mathrm{B}}T)dT \tag{1.43}$$

by making use of the definition of partition function given in equation (1.14). Then we immediately obtain

$$
\begin{aligned}
\frac{k_{\mathrm{B}}N}{\mathcal{Z}}d\mathcal{Z} &= -\frac{1}{T}\sum_i \underbrace{\frac{N}{\mathcal{Z}}p_i \exp(-E_i/k_{\mathrm{B}}T)}_{=\,n_i}dE_i + \frac{1}{T^2}\sum_i \underbrace{\frac{N}{\mathcal{Z}}p_i \exp(-E_i/k_{\mathrm{B}}T)E_i}_{=\,n_i}\,dT \\
&= -\frac{1}{T}\sum_i n_i dE_i + \frac{1}{T^2}\left(\sum_i n_i E_i\right)dT
\end{aligned}
\tag{1.44}
$$

where we recalled the Boltzmann distribution law given in equation (1.15).

In order to proceed, let us consider the most general expression for the variation of the total internal energy of the system occurring during the transformation. From equation (1.1) it is easily calculated

$$d\mathcal{U} = \sum_i n_i dE_i + \sum_{i=1}^{N} E_i dn_i \tag{1.45}$$

which represents the microscopic counterpart of the first law of thermodynamics stated in equation (1.36) and here referred to the case in which the total number of particles is not varied since, as commented above, the system is isolated (no contribution from chemical work). Consistently with the macroscopic version of this energy balance principle, we can attach to the two contributions a rather different physical origin: they, respectively, describe how the internal energy is affected (i) as the energies of single-particle states vary by dE_i and (ii) as their partition is modified by a particle redistribution dn_i. We conclude that $dW = -\sum_i n_i\, dE_i$ is the work done by the system upon a volume variation; the negative sign indicates that such a work is performed at the expense of the internal energy of the system, as sketched in figure 1.3 in the simple case of a one-dimensional assembly of free particles confined within a potential box whose width is enlarged. On the other hand, it is natural to identify $dQ = \sum_i E_i dn_i$ as the exchanged heat.

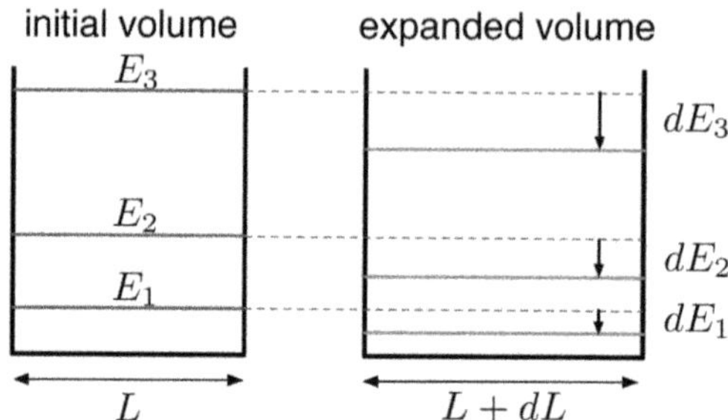

Figure 1.3. Pictorial representation of the change in the energy levels dE_i upon volume variation.

Using these results, equation (1.44) is finally recast in the following form

$$\frac{k_{\mathrm{B}}N}{\mathcal{Z}}d\mathcal{Z} = \frac{1}{T}\,dW + \frac{\mathcal{U}}{T^2}\,dT \tag{1.46}$$

which, if inserted into equation (1.42), immediately leads to the result we were looking for

$$dS = \frac{1}{T}(d\mathcal{U} + dW) = \frac{1}{T}d\mathcal{Q} \tag{1.47}$$

in virtue of the first law stated in equation (1.36).

In conclusion, this journey back-and-forth through thermodynamics and statistical physics provides clear evidence of how much it pays to develop an atomistic description of physical systems: phenomenological definitions introduced at the macroscopic level end up finding a satisfactory microscopic justification based on first principles.

1.6.3 Entropy and ignorance

As extensively discussed above, once a macrostate is assigned, the system can occupy a (possibly very large) variety of microstates. Physically, this reflects in the fact that *a macroscopic measurement cannot determine the actual microstate*. Let us consider a model system whose N particles can occupy just two discrete energy levels E_1 and E_2. As sketched in figure 1.4 the macrostate with energy $\mathcal{U} = NE_1$ has only one microstate since it corresponds to the sole partition with all particles on the ground state with energy E_1 (top panel of figure 1.4); on the other hand, if we consider the macrostate with energy $\mathcal{U} = (N - 1)E_1 + E_2$ we have N different choices to select the particle to place on the higher-energy level, corresponding to as many different partitions (middle panels of figure 1.4); next, the macrostate with energy $\mathcal{U} = (N - 2)E_1 + 2E_2$ corresponds to $N(N - 1)/2$ different microstates, corresponding to the number of pairs we can select out of the set of N particles (bottom panels of figure 1.4); and so on. We immediately realise that *the full knowledge of the macrostate corresponds to different levels of ignorance about its compatible microstates*: in the first case we have no ignorance (just one microstate is possible), in the

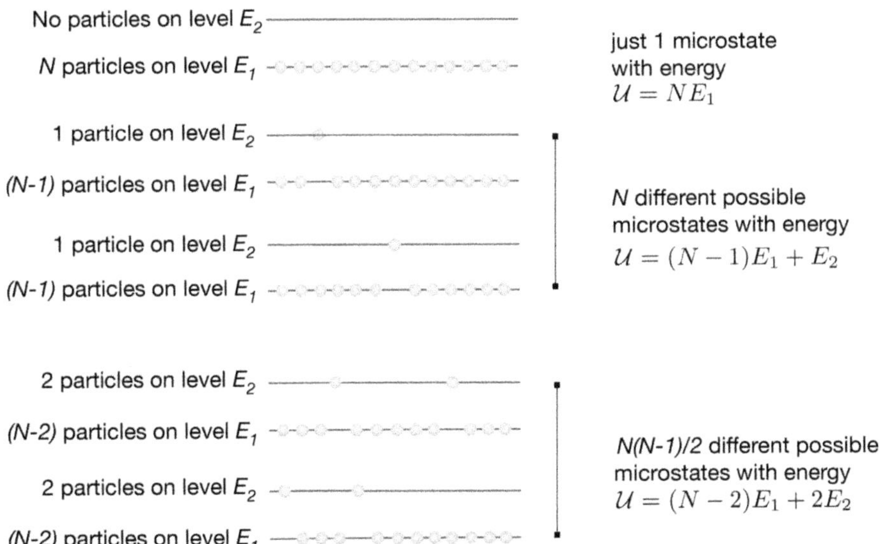

Figure 1.4. Pictorial representation of the microstate with $\mathcal{U} = NE_1$ (top) and some microstates with $\mathcal{U} = (N-1)E_1 + E_2$ (middle) and $\mathcal{U} = (N-2)E_1 + 2E_2$ (bottom).

second case we have some ignorance, in the third case we suffer an even worse ignorance (by constitutive hypothesis N is large and, therefore, $N(N-1)/2 > N$), and so on. This argument suggests that we can try to link our ignorance about the system microstate to its entropy.

Let us start with the assumption that *all microstates corresponding to the same macrostate are equally likely*[16]. This assumption is physically quite reasonable: there is no reason to expect microstates with equal macroscopic properties to have different probability of occurrence. While in appendix B this concept will be critically readdressed, leading to an alternative definition of entropy due to Gibbs, here we will rely on such an assumption. Next, let us introduce a function quantifying our ignorance of the microstate, as defined above. For reasons to be clear in a while, we will name this function S; since all accessible microstates are equally likely, S can only be a function of their number Ω; formally, $S = S(\Omega)$. This function has some intriguing properties, namely: (i) whenever we have no ignorance, that is whenever $\Omega = 1$, it must be zero, or equivalently, we can set $S(\Omega = 1) = 0$; (ii) by considering two macrostates a and b with $\Omega_a > \Omega_b$ we set $S(\Omega_a) > S(\Omega_b)$ since we suffer more ignorance for the first system; (iii) finally, if we consider two independent systems Σ_1 and Σ_2, respectively, with a number of possible microstates Ω_{Σ_1} and Ω_{Σ_2}, then the total ignorance of the joint system $\Sigma_1 + \Sigma_2$ is the sum of the ignorances we have about its separated components and formally we write

---

[16] We remark that microstates corresponding to *different* macrostates are not constrained to such a same-likelihood condition.

$S(\Omega_{\Sigma_1+\Sigma_2}) = S(\Omega_{\Sigma_1}\Omega_{\Sigma_2}) = S(\Omega_{\Sigma_1}) + S(\Omega_{\Sigma_2})$. In summary, this ignorance function S must obey the following properties

$$S(1) = 0$$
$$S(xy) = S(x) + S(y) \tag{1.48}$$
$$\text{if } x > y \rightarrow S(x) > S(y)$$

which are fulfilled by the function logarithm. Therefore, we straightforwardly set

$$S(x) = k \ln x + \text{constant} \tag{1.49}$$

that, since $S(1) = 0$, further reduces to $S(x) = k \ln x$. If we now set $k = k_B$ we eventually obtain equation (1.37). This formal reasoning completes the plausibility argument developed in the previous section to define entropy by means of equation (1.6.2) which is now linked to the content of information (or, reversely, to the degree of ignorance) we have about the system.

The conceptual link we just established between the entropy and the ignorance we have about the system allows for an interesting excursion in the realm of information theory. Here the key ingredients are not physical particles, but rather 'information quanta' (usually referred to as *bits*), which are restricted to just two values, namely 0 and 1. If we have N bits, we have 2^N possible information states which correspond to a physical entropy $S_{\text{init}} = Nk_B \ln 2$. If we want to erase the information stored in a computational device, we should set all bits to the same value (it is irrelevant whether 0 or 1): a situation which corresponds to $S_{\text{final}} = 0$. Erasing the stored information requires therefore an entropy variation $\Delta S = S_{\text{final}} - S_{\text{init}} = -Nk_B \ln 2$. As discussed in section 1.6.2, the second law of thermodynamics imposes that the entropy change must always be non-negative and, therefore, we conclude that we need to increase the environment entropy by an amount $k_B \ln 2$ (sometimes referred to as Shannon entropy) for any bit erased in the device memory. This corresponds to a minimum $k_B T \ln 2$ dissipated heat per erased bit. Deleting a file on our personal computer does increase the environment temperature!

1.7 The ideal monoatomic gas

The ideal monoatomic gas is a simple model system made by *identical, structureless, non-interacting particles*. This implies that the energy of each particle is just translational kinetic, since no interactions among them are present nor must internal roto-vibrational contributions be taken into account.

Although we are addressing an idealised physical situation, the results obtained for the ideal monoatomic gas are largely valid for real dilute gases (kept at intermediate or high temperature). More importantly, this model represents a benchmark system where calculations can be exploited analytically and the results can be directly compared with similar ones obtained by other theoretical approaches, like for instance the kinetic theory of gases [3].

In short, the physics of the ideal monoatomic gas provides paradigmatic results which exceed the limits of the model for relevance and general validity.

1.7.1 Partition function

Let us consider an assembly of N independent (that is, free) particles confined in a volume V. It is a well known fundamental result of quantum mechanics that the energy spectrum of a confined particle is discrete [12–14]; however, the spacing between two adjacent levels is inversely proportional to $V^{2/3}$. This means that for a large enough volume, successive levels are so close that their spectrum can be treated as continuous to a very good approximation. This is precisely the case we are treating: a gas of pointlike particles contained within a macroscopic volume.

The fact that the energy spectrum is continuous has a twofold implication. First of all, we can profitably make use of the concept of *density of energy levels* $G(E)$ for the confined particles, where it is understood that $G(E)\,dE$ provides the number of levels in the energy interval $[E, E + dE]$; a standard quantum mechanical calculation [12–14] proves that

$$G(E) = \frac{4\pi V}{h^3}(2m^3 E)^{1/2} \tag{1.50}$$

where $h = 4.135\,7 \times 10^{-15}$ eV s is the Planck constant and m is the particle mass. Next, we proceed by replacing the partition function given in equation (1.14) with its continuous counterpart

$$\mathcal{Z} = \int_0^{+\infty} \exp(-E/k_\mathrm{B}T)G(E)dE \tag{1.51}$$

where in setting the limits of the energy integral we have duly taken into account that in the present case E is just kinetic energy and, therefore, is a positive quantity. By replacing equation (1.50) in equation (1.51) we easily get the *partition function of the ideal monoatomic gas*

$$\mathcal{Z} = \frac{V}{h^3}(2\pi m k_\mathrm{B}T)^{3/2} \tag{1.52}$$

where for the calculation we used appendix A reporting some integrals useful in statistical physics. This result shows that *the partition function of the gas depends both on its volume and temperature.*

1.7.2 The equation of state

In applying the general expression for the variation of the partition function given in equation (1.46), we must take into account that for an ideal gas the only form of mechanical work is related to volume variations or, more formally, $dW = P\,dV$ where P is the system pressure. This allows us to recast equation (1.46) in the form

$$\frac{k_\mathrm{B}N}{\mathcal{Z}}d\mathcal{Z} = \frac{P}{T}dV + \frac{\mathcal{U}}{T^2}dT \tag{1.53}$$

which, in the case of a constant-temperature process, allows us to find the *statistical definition of pressure*

$$P = k_{\mathrm{B}}NT\frac{\partial \ln \mathcal{Z}}{\partial V}\bigg|_{T} \tag{1.54}$$

in the form of a relationship between temperature, pressure, and partition function. By inserting the explicit form of $\mathcal{Z}$ provided in equation (1.52) we immediately find

$$PV = k_{\mathrm{B}}NT = nRT \tag{1.55}$$

a result known as *equation of state for the ideal monoatomic gas*; as is customary, the number of particles has been set $N = n\mathcal{N}_{\mathrm{A}}$ where $\mathcal{N}_{\mathrm{A}} = 6.022 \times 10^{23}$ mol^{-1} is the Avogadro number, corresponding to the number of particles in one mole of gas.

Through equation (1.55) we can define the experimental value of the Boltzmann constant anticipated in equation (1.28). Following the standard convention, we define as *the normal condition* the equilibrium of a mixture of liquid water and solid water (ice) at the standard pressure of 1 atm. The corresponding temperature is assigned a conventional value T_0 by choosing a convenient scale. If we now bring a gas in contact with the mixture and wait until it reaches the equilibrium, we can measure its pressure P_0 and volume V_0. The Boltzmann constant is accordingly defined as $k_{\mathrm{B}} = P_0 V_0 / N T_0$ and its numerical value is straightforwardly obtained once we know the mass of each atom forming the gas.

1.7.3 Energy and heat capacity

In order to calculate the total internal energy $\mathcal{U}$ of an ideal gas in equilibrium condition we make use of equation (1.33) where the partition function is provided by equation (1.52)

$$\langle E \rangle = k_{\mathrm{B}}T^2 \frac{d}{dT}\ln \mathcal{Z} = \frac{3}{2}k_{\mathrm{B}}T \tag{1.56}$$

from which we immediately obtain

$$\mathcal{U} = \frac{3}{2}Nk_{\mathrm{B}}T \tag{1.57}$$

a result which can be alternatively derived within the kinetic theory [3]. A different useful formulation is obtained by considering n moles of ideal gas in equilibrium at temperature T: equation (1.57) easily leads to

$$\mathcal{U} = \frac{3}{2}nRT \tag{1.58}$$

where $R = k_{\mathrm{B}}\mathcal{N}_{\mathrm{A}} = 8.313\,4$ J K^{-1} mol^{-1} is the universal gas constant. This very popular result, consistent with classical macroscopic thermodynamics [8–10], states that *the internal energy on an ideal gas in equilibrium depends only on its temperature*. From this result we obtain immediately the *molar constant-volume heat capacity* C_{V} of the ideal monoatomic gas.

$$C_{\mathrm{V}} = \frac{1}{n}\frac{\partial \mathcal{U}}{\partial T}\bigg|_{\mathrm{V}} = \frac{3}{2}R \tag{1.59}$$

The constant-pressure counterpart of this quantity is calculated from the system enthalpy $\mathcal{H} = \mathcal{U} + PV$, which as outlined in appendix C is a thermodynamical potential representing the work available when the system is kept in a $P = $ constant condition. For the monoatomic ideal gas we have

$$\mathcal{H} = \frac{3}{2}nRT + PV = \frac{3}{2}nRT + nRT = \frac{5}{2}nRT \tag{1.60}$$

and therefore we can define the *molar constant-pressure heat capacity* C_P as

$$C_P = \frac{1}{n}\frac{\partial \mathcal{H}}{\partial T}\bigg|_P = \frac{5}{2}R \tag{1.61}$$

a result immediately leading to

$$C_P - C_V = R \tag{1.62}$$

which is known as the *Meyer relation for the ideal gas*. This equation states that the amount of heat required to increase the temperature by one degree Kelvin of an ideal gas at constant pressure is greater than that required to achieve the same increase at constant volume. The physical reason is that the heat supplied to the system in the first case is divided between temperature increase and mechanical work. More precisely, the work per mole produced at constant pressure is

$$W = \frac{1}{n}\int PdV = \frac{1}{n}\int_T^{T+1} nRdT = R \tag{1.63}$$

for a $\Delta T = 1$ K temperature increase.

1.7.4 The Maxwell distribution law

By means of equation (1.56), we can look at the superior Boltzmann theory as the statistical foundation of the kinetic theory of gases[17]. In order to further develop this concept, let us consider the Boltzmann distribution law given in equation (1.15) and replace the occupation probabilities p_i with their continuous counterpart $G(E)dE$; the resulting *number dN of particles with energy in the range* $[E, E + dE]$ is given by

$$dN = \frac{N}{\mathcal{Z}}\exp(-E/k_BT)G(E)dE = \frac{N}{\mathcal{Z}}\frac{4\pi V}{h^3}(2m^3)^{1/2}E^{1/2}\exp(-E/k_BT)dE \tag{1.64}$$

which, by using equation (1.52) for the equilibrium partition function, immediately leads to

---

[17] Incidentally, we remark that this result suggested the identification anticipated in equation (1.19). The kinetic theory (developed prior to statistical mechanics) already proved that $\langle E \rangle = 3k_BT/2$ on the basis of purely mechanical arguments; on the other hand, the Boltzmann statistics without equation (1.19) would lead to the formulation $\langle E \rangle = 3/2\beta$. It is then straightforward to identify the Lagrange multiplier as $\beta = 1/k_BT$.

$$\frac{dN}{dE} = \frac{2\pi N}{(\pi k_B T)^{3/2}} E^{1/2} \exp(-E/k_B T) \tag{1.65}$$

a result known as the *Maxwell energy distribution law* in an ideal gas. Since in the case under study the particle energy is purely kinetic, the energy only depends on the velocity v and we have

$$\frac{dN}{dv} = \frac{dN}{dE}\frac{dE}{dv} = \frac{dN}{dE}mv \tag{1.66}$$

where we made use of equation (1.65) by setting $E = mv^2/2$. By combining these intermediate results we obtain

$$\frac{dN}{dv} = 4\pi N\left(\frac{m}{2\pi k_B T}\right)^{3/2} v^2 \exp(-mv^2/2k_B T) \tag{1.67}$$

which represents the *Maxwell velocity distribution law* in an ideal gas, providing the number of particles with velocity in the range $[v, v + dv]$ (irrespective of the direction of motion) when the system is in equilibrium at temperature T.

The two Maxwell laws are pretty well confirmed by direct experimental results. Let us consider the experimental setup sketched in figure 1.5: the particles of a monoatomic gas kept at controlled temperature by a thermostat are effused through a nozzle placed on the surface of the containing box and collimated to form a thin beam. The beam is directed towards two slotted disks which rotate around their common axis at constant angular frequency ω; their slots are displaced by an angle θ, while the centres of the two disks are placed at distance s. The particles can pass through the two-disk chopper only if they propagate with translational velocity $v = s\omega/\theta$, since only in this case can the beam pass through both slots, one after the other; in this condition, they are collected by a detector. Therefore, by tuning ω or θ

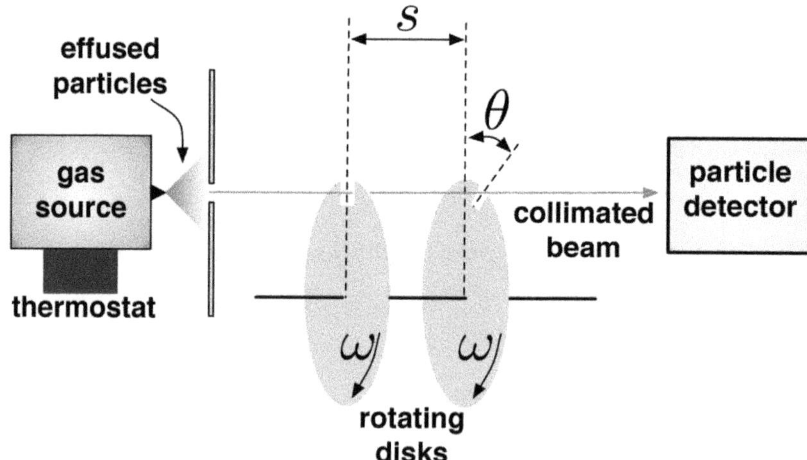

Figure 1.5. Experimental setup to determine the velocity distribution within a gas.

we can effectively select the velocity of the transmitted particles and count their number. The experiment consists in fixing the thermostat temperature and counting the number of transmitted particles at several different velocities. This provides the velocity distribution at that temperature, just in perfect agreement with equation (1.67). By now changing the gas temperature and repeating the measure, we can experimentally determine how the velocity distribution depends on T; once again, empirical data confirm the theoretical prediction by the Maxwell law[18].

The Maxwell laws are not monotonic: thus, it is meaningful to calculate *the most probable energy* E_{mp} *and velocity* v_{mp} for the particles in the ideal monoatomic gas. They are obtained by looking at the maximum of the distributions provided by equations (1.65) and (1.67)

$$E_{mp} = \frac{1}{2}k_BT \quad \text{and} \quad v_{mp} = \sqrt{\frac{2k_BT}{m}} \tag{1.68}$$

and turn out to only depend on temperature. Similarly, we can calculate *the average velocity* $\langle v \rangle$ of the particles. We simply apply the definition of average

$$\langle v \rangle = \frac{1}{N}\int_0^{+\infty} v\frac{dN}{dv}dv = 4\pi N\left(\frac{m}{2\pi k_BT}\right)^{3/2}\int_0^{+\infty} v^3 \exp(-mv^2/2k_BT)dv \tag{1.69}$$

where we used equation (1.67) for the term dN/dv. By integration by parts we obtain

$$\langle v \rangle = \sqrt{\frac{8k_BT}{\pi m}} \tag{1.70}$$

proving that the average velocity is always larger that the most probable one. We can also evaluate *the mean square velocity* $\langle v^2 \rangle$ of the particles by a similar averaging procedure

$$\langle v^2 \rangle = \frac{1}{N}\int_0^{+\infty} v^2\frac{dN}{dv}dv \tag{1.71}$$

but in this case we can transform the integral as

$$\langle v^2 \rangle = \frac{2}{m}\frac{1}{N}\int_0^{+\infty} EdN = \frac{2}{m}\langle E \rangle \tag{1.72}$$

since $v^2 = 2E/m$. We eventually get

---

[18] A more indirect way to verify equation (1.65) is to consider a chemical reaction which can only occur by involving particles with energy above a known value E_{react}. Accordingly, the rate of this chemical reaction at a given temperature depends on the number of particles whose energy exceeds E_{react}. Now, we can measure the reaction rate at different temperatures and we can estimate by equation (1.65) the number of active particles at the same temperatures with energy higher than E_{react}. The key point is that if the observed rates correlated with the estimated particles number at each temperature, we could definitely validate the Maxwell law for the energy distribution. Good enough, this is indeed the case: data correlate and, therefore, the empirical evidence does support statistical theory predictions.

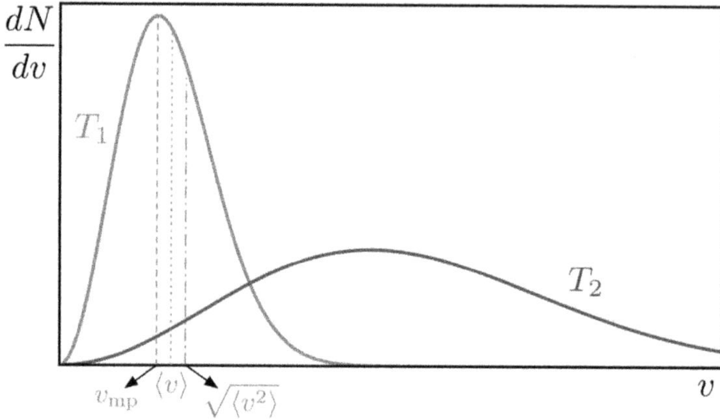

Figure 1.6. Maxwell law for the velocity distribution (see equation (1.67)) calculated for the same system at temperature T_1 (red line) and T_2 (blue line) with $T_1 < T_2$. In the case of lower temperature, the most probable v_{mp} velocity, the average velocity $\langle v \rangle = 1.13 v_{mp}$, and the (square root of the) mean square velocity $\sqrt{\langle v^2 \rangle} = 1.15 v_{mp}$ are, respectively, indicated by a dashed, dotted, and dash-dotted line.

$$\langle v^2 \rangle = \frac{3k_{\mathrm{B}}T}{m} \tag{1.73}$$

where equation (1.56) has been used. We further calculate

$$\sqrt{\langle v^2 \rangle} = \sqrt{\frac{3k_{\mathrm{B}}T}{m}} \tag{1.74}$$

which allows us to compare this value with the average and most probable velocities, as reported in figure 1.6 where we have made use of the numerical estimations $\langle v \rangle \simeq 1.13\, v_{mp}$ and $\sqrt{\langle v^2 \rangle} \simeq 1.25\, v_{mp}$.

1.7.5 The entropy state function

As a matter of fact, for a monoatomic system three different equations of state $T = T(S, V, N)$, $P = P(S, V, N)$, and $\mu = \mu(S, V, N)$ can be formulated: equation (1.55) is just one of them. It must be stressed, however, that each single of such relations does not imply a complete knowledge of the thermodynamical properties of the system: this is only provided by the full set of them or, alternatively, by the single entropy state function. It is equally possible to formulate the thermodynamical theory by making use of the internal energy, which is as well a state function. We will, respectively, refer to the 'entropy representation' or to the 'energy representation' [8]. In this section we are calculating the entropy state function.

By inserting the partition function of the ideal monoatomic gas given in equation (1.52) into the general expression for entropy provided through equation (1.40) or (1.41) we get

$$S = \frac{5}{2}k_{\mathrm{B}}N + k_{\mathrm{B}}N \ln \frac{V(2\pi m k_{\mathrm{B}}T)^{3/2}}{Nh^3} \tag{1.75}$$

a result referred to as the *Sackur–Tedrode state equation*. We remark that during the early stages of development of the statistical approach it has been cumbersome to elaborate this result since the subtleties underlying the concept of distinguishable or indistinguishable particles were not all that clear. The discussion that arose around this problem and the apparent contradictions of the theory were referred to as the 'Gibbs paradox' [3, 4].

In order to provide an example of the powerfulness of the entropy state equation, let us consider an *adiabatic process*, namely a process not involving any exchange of heat between the system and its environment (we will also assume a constant mass). We assume that the ideal monoatomic gas undergoes an expansion from the initial volume V_{init} to the final one V_{fin}; its temperature accordingly varies from T_{init} to T_{fin}. By calculating explicitly the corresponding entropy values S_{init} and S_{fin} and setting $\Delta S = S_{fin} - S_{init} = 0$ we obtain

$$\ln V_{init} + \frac{3}{2} \ln T_{init} = \ln V_{fin} + \frac{3}{2} \ln T_{fin} \tag{1.76}$$

and therefore

$$VT^{3/2} = \text{constant} \tag{1.77}$$

which is the equation describing an adiabatic transformation, as phenomenologically found in macroscopic thermodynamics [8, 9].

1.8 On the statistical concept of irreversible process

As a final argument of this tutorial approach to the statistical description of a classical system, we will address the very concept of *irreversible process*[19]. To this aim, we consider an ideal monoatomic gas initially confined within a container with volume V_{init}. Let us also suppose that such a container is linked through a removable wall to a second *empty* container m-times larger, i.e. with volume $V_{final} = mV_{init}$. By some means we now remove the separating wall and, therefore, we observe a *free expansion* of the gas from the initial volume V_{init} to the final one V_{final}, as pictorially shown in figure 1.7. During this expansion the gas temperature does not change, because the internal energy of the gas does not vary (see equation (1.57)): simply, the

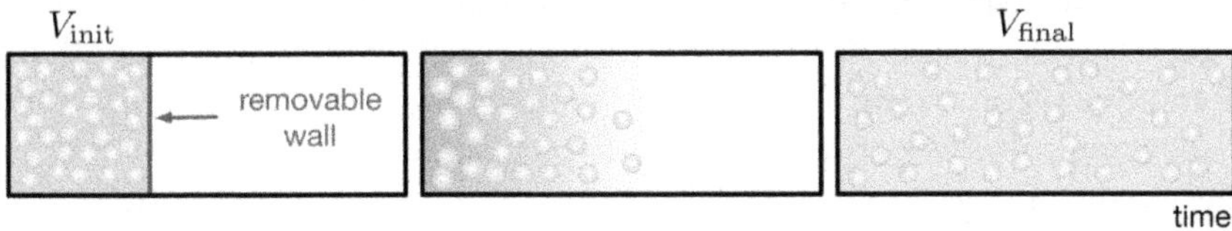

Figure 1.7. Pictorial representation of the free expansion process of a ideal monoatomic gas from the initial volume V_{init} to the final one V_{final}.

---

[19] We recall that in section 1.6 we basically proclaimed as *reversible* any process occurring quasi-statically, that is displaying over a time scale much larger than the intrinsic relaxation time of the system.

atoms move in a larger space and the gas does not need to spend work in expanding. From the macroscopic point of view this quintessentially is an irreversible process: no one reasonably expects that the gas, without external actions, can spontaneously return to the initial condition in which it occupied only the volume V_{init}. From the statistical point of view, the interpretation is more subtle and it is fully enlightened by proceeding with the calculation of the *entropy variation for a free expansion*.

Let us consider equation (1.75) and introduce the shortcut notation

$$\bar{V} = \frac{V}{N} \quad \text{and} \quad \bar{S} = \frac{5}{2} k_B N + k_B N \ln \frac{(2\pi m k_B)^{3/2}}{h^3} \tag{1.78}$$

which allows us to calculate the gas entropy in the initial and final conditions as

$$S_{init} = k_B N \ln \left(\bar{V}_{init} T^{3/2} \right) + \bar{S} \quad \text{and} \quad S_{final} = k_B N \ln \left(\bar{V}_{final} T^{3/2} \right) + \bar{S} \tag{1.79}$$

and the corresponding *entropy variation for the free expansion* as

$$\Delta S = S_{final} - S_{init} = k_B N \ln \left(\frac{\bar{V}_{final}}{\bar{V}_{init}} \right) = k_B N \ln m > 0 \tag{1.80}$$

since $m > 1$. This result ensures that the statistical description of the free expansion process is compatible with its thermodynamical counterpart: following whichever of the two approaches, the variation of the system entropy for a free expansion is consistently found to be positive. However, by remembering equation (1.37) we can recast equation (1.80) in the different form

$$\Delta S = k_B \ln \frac{\Omega_{final}}{\Omega_{init}} \tag{1.81}$$

which leads immediately to

$$\ln \frac{\Omega_{final}}{\Omega_{init}} = N \ln m \quad \rightarrow \quad \frac{\Omega_{final}}{\Omega_{init}} = m^N \tag{1.82}$$

where we exploited the link between the value of the entropy of any given macrostate and the corresponding number of compatible microstates. In general N is very large and, therefore, $\Omega_{final} \gg \Omega_{init}$ that is: *the final state is much more probable than the initial one*. On the other hand, if the reverse process $V_{final} \rightarrow V_{init}$ occurred, then we would get $\Omega_{init}/\Omega_{final} = m^{-N}$ which corresponds to an extremely small quantity indeed! In conclusion, *from the statistical point of view the free compression is not strictly speaking forbidden, rather it is 'just' very unlikely.*

While we developed this concept by studying a very simple model system, the conclusion is valid in general: *irreversible processes are processes that very unlikely occur in the opposite direction.* However, the smaller the number of particles forming the system, the higher is the probability to observe a process to occur in both directions. In this case, however, the statistical approach lacks a fundamental constitutive hypothesis (namely, that we are dealing with large systems) and, therefore, it is more fruitful to follow a mechanical one.

References

[1] Demtröder W 2010 *Atoms, Molecules and Photons* (Berlin: Springer)

[2] Colombo L 2021 *Solid State Physics: A Primer* (Bristol: IOP Publishing)

[3] Reif F 1987 *Fundamentals of Statistical and Thermal Physics* (New York: McGraw-Hill)

[4] Kennett M P 2021 *Essential Statistical Mechanics* (Cambridge: Cambridge University Press)

[5] Chandler D 1987 *Introduction to Modern Statistical Mechanics* (Oxford: Oxford University Press)

[6] Frenkel D and Smit B 1996 *Understanding Molecular Simulations: From Algorithms to Applications* (San Diego, CA: Academic)

[7] Lebon G, Jou D and Casas-Vázquez J 2008 *Understanding Non-equilibrium Thermodynamics* (Berlin: Springer)

[8] Callen H 1985 *Thermodynamics and an Introduction to Thermostatistics* (New York: Wiley)

[9] Dittman R and Zemansky M 1997 *Heat and Thermodynamics* 7th edn (New York: McGraw-Hill)

[10] Swendsen R H 2012 *An Introduction to Statistical Mechanics and Thermodynamics* (Oxford: Oxford University Press)

[11] Glazer M and Wark J 2001 *Statistical Mechanics: A Survival Guide* (Oxford: Oxford University Press)

[12] Eisberg R and Resnick R 1985 *Quantum Physics of Atoms, Molecules, Solids, Nuclei, and Particles* 2nd edn (Hoboken, NJ: Wiley)

[13] Miller D A B 2008 *Quantum Mechanics for Scientists and Engineers* (New York: Cambridge University Press)

[14] Bransden B H and Joachain C J 2000 *Quantum Mechanics* (Englewood Cliffs, NJ: Prentice-Hall)

IOP Publishing

Statistical Physics of Condensed Matter Systems
A primer
Luciano Colombo

Chapter 2

Thermal properties of classical gases

Syllabus—We apply the statistical theory developed in the previous chapter to investigate the thermal properties of gases. We consider at first the case of an ideal polyatomic gas, whose particles have both rotational and vibrational internal degrees of freedom, and we work out a general expression for its molar heat capacity. We make use of this investigation to elaborate the fundamental principle of equipartition of energy, indeed a cornerstone of the statistical physics of classical systems. Next, we extend our treatment to real gases, for which a suitable equation of state is elaborated. Eventually, as a further example of how the statistical approach works, we investigate the magnetisation and polarisation phenomena under the action of an external magnetic and electric field, respectively.

2.1 Thermal properties of polyatomic ideal gases

In section 1.7 we investigated the thermal properties of a gas made by structureless and non-interacting particles. Here, by releasing the first assumption, we admit that the elementary constituents of the systems are indeed molecules with *rotational* and *vibrational* internal degrees of freedom, in addition to the translational ones already included in the previous model. The energy of each molecule is therefore the sum of four contributions, referred to as electronic, translational, rotational and vibrational energy. The thermal bath is usually too weak to excite transitions between electronic states[1] and, therefore, we will neglect the electronic energy in the following analysis

---

[1] While electronic transitions require an energy of the order of $\sim$1 eV, at room temperature the thermal bath has an energy content of $k_B T_{room} = 0.002\,5$ eV. In order to thermally excite electrons, we need a temperature as large as several thousands degree Kelvin: at such a high temperature the molecule is likely to be dissociated by inter-molecular collisions.

since it only provides a contribution which, to a very good extent, may be considered constant. On the other hand, molecular rotations and vibrations are predicted and observed to fall in the 10^{-4} eV and 10^{-3} eV energy range, respectively [1, 2]: such degrees of freedom can be thermally excited, even at relatively low temperatures; therefore, they must be duly taken into account, as well as the translational ones which are treated exactly as in the case of the monoatomic ideal case. In other words, the translational energy describes the drift motion of the molecular centre of mass and it is calculated as $\mathcal{U}_{tr} = 3Nk_BT/2 = 3nRT/2$, where in this case N and n are, respectively, the number of molecules and the number of moles forming the gas system. More intriguing is the case of rotational and vibrational contributions, both requiring a full quantum treatment which will be here developed for the simple case of diatomic molecules.

2.1.1 Rotational energy

Elementary quantum mechanics provides the rotational energy for a diatomic molecule in the form [2]

$$E_{rot} = Bhc\, r(r + 1) \tag{2.1}$$

where $r = 0, 1, 2, 3, \ldots$ is the *rotational quantum number*, while $B = \hbar/4\pi Ic$ and I is the molecular moment of inertia. Each rotational level has degeneracy $p_r = 2r + 1$. According to the Boltzmann distribution law derived in section 1.3, the *number n_r of molecules in the r-th rotational state* when the gas is in equilibrium at the temperature T is

$$n_r = \frac{N}{\mathscr{Z}_{rot}} p_r \exp(-E_{rot}/k_BT) \tag{2.2}$$

where

$$\mathscr{Z}_{rot} = \sum_r (2r + 1)\exp[-Bhcr(r + 1)/k_BT] \tag{2.3}$$

is the *rotational partition function*. We easily get the most general expression for the *rotational internal energy of the gas* in the form

$$\mathcal{U}_{rot} = k_BNT^2\frac{d}{dT}\ln \mathscr{Z}_{rot} = nRT^2\frac{d}{dT}\ln \mathscr{Z}_{rot} \tag{2.4}$$

according to equation (1.32); the last equality holds if $N = n\mathcal{N}_A$.

2.1.2 Vibrational energy

Under the harmonic approximation [2], the vibrational motion of a diatomic molecule corresponds to a one-dimensional quantum oscillator with energy

$$E_{vib} = \left(\lambda + \frac{1}{2}\right)\hbar\omega_0 \tag{2.5}$$

where $\lambda = 0, 1, 2, 3, \ldots$ is the *vibrational quantum number* and $\omega_0 = \sqrt{\gamma/M}$ is the fundamental oscillation frequency which, in turn, depends on the molecular reduced mass M and the force constant $\gamma = [d^2 V(R)/dR]_{R_0}$, where $V(R)$ is the interatomic potential and R_0 is the interatomic equilibrium distance of the molecule.

Vibrational energy levels are non-degenerate and, therefore, their intrinsic occupation probability is $p_\lambda = 1$ for any possible value of the quantum number λ. The Boltzmann distribution law derived in section 1.3 leads to predicting the *number n_λ of molecules in the λ-th vibrational state* when the gas is in equilibrium at temperature T as

$$n_\lambda = \frac{N}{\mathcal{Z}_{\text{vib}}} \exp(-E_{\text{vib}}/k_{\text{B}}T) \tag{2.6}$$

where

$$\mathcal{Z}_{\text{vib}} = \sum_{\lambda} \exp\left[(\lambda + 1/2)\hbar\omega_0/k_{\text{B}}T\right] \tag{2.7}$$

is the *vibrational partition function*. It is convenient to introduce the *characteristic vibrational temperature* $T_{\text{vib}} = \hbar\omega_0/k_{\text{B}}$ so that equation (2.7) can be recast in a more convenient form[2]

$$\mathcal{Z}_{\text{vib}} = \exp(-T_{\text{vib}}/2T) \sum_{v} \exp(-\lambda T_{\text{vib}}/T) = \frac{\exp(-T_{\text{vib}}/2T)}{1 - \exp(-T_{\text{vib}}/T)} \tag{2.8}$$

from which we straightforwardly calculate the most general expression for *vibrational internal energy of the gas* in the form

$$\mathcal{U}_{\text{vib}} = k_{\text{B}}NT^2 \frac{d}{dT} \ln \mathcal{Z}_{\text{vib}} = nRT^2 \frac{d}{dT} \ln \mathcal{Z}_{\text{vib}} \tag{2.9}$$

according to equation (1.32); the last equalty holds if $N = n\mathcal{N}_{\text{A}}$.

2.1.3 Molar constant-volume heat capacity

The total internal energy $\mathcal{U}$ of a diatomic molecular gas (under the assumption of neglecting the electronic contribution, as commented above) can be written as

$$\mathcal{U} = \mathcal{U}_{\text{tr}} + \mathcal{U}_{\text{rot}} + \mathcal{U}_{\text{vib}} \tag{2.10}$$

from which we immediately calculate its *molar constant-volume heat capacity* as

$$C_V = \frac{1}{n}\left(\frac{\mathcal{U}_{\text{tr}}}{\partial T}\bigg|_V + \frac{\mathcal{U}_{\text{rot}}}{\partial T}\bigg|_V + \frac{\mathcal{U}_{\text{vib}}}{\partial T}\bigg|_V \right) \tag{2.11}$$

where the first term is easily computed as

---

[2] Just put $x = \exp(-T_{\text{vib}}/T)$ and make use of the decreasing geometric progression $\sum_n x^n = 1/(1 - x)$ which holds for any $x < 1$ as is indeed the case since $T_{\text{vib}}/T > 0$ in any case.

$$\frac{1}{n}\frac{\partial\, \mathcal{U}_{\text{tr}}}{\partial T}\bigg|_V = \frac{3}{2}R \tag{2.12}$$

while the rotational and vibrational contributions must be handled with some care.

Let us first consider the rotational term. The energy scale of rotations and vibrations is, respectively, set by the Bhc term appearing in equation (2.1), and by the $\hbar\omega_0$ term appearing in equation (2.5). In any case, it is found that $Bhc \ll \hbar\omega_0$ [2]. Accordingly, by defining the *rotational characteristic temperature* $T_{\text{rot}} = Bhc/k_B$ we can argue that for any $T > T_{\text{rot}}$ a very large number of rotational levels is occupied. This state of affairs is tantamount to saying that the summation appearing in equation (2.3) can be replaced by an integral

$$\mathcal{Z}_{\text{rot}} = \int_0^{+\infty} 2r\,\exp\big[-(T_{\text{rot}}/T)r^2\big]dr = T/T_{\text{rot}} \tag{2.13}$$

where by treating the r variable as a continuous one we have replaced $(2r + 1) \to 2r$ and $r(r + 1) \to r^2$ with no loss of accuracy since r is large. This result allows us to write the rotational energy given in equation (2.4) in a more simple form

$$\mathcal{U}_{\text{rot}} = nRT^2\frac{d}{dT}\left(\ln\frac{T}{T_{\text{rot}}}\right) = nRT \tag{2.14}$$

which is understood to be valid whenever $T > T_{\text{rot}}$.

We now move to consider the vibrational term. By inserting equation (2.8) into equation (2.9) we obtain

$$\mathcal{U}_{\text{vib}} = nRT_{\text{vib}}\left[\frac{1}{2} + \frac{1}{\exp(T_{\text{vib}}/T) - 1}\right] \tag{2.15}$$

where the first term in the square parenthesis corresponds to the zero-point motion and it is irrelevant for any thermal process where only energy differences count. If the temperature is large enough, the T_{vib}/T is small and we can therefore expand the exponential term $\exp(T_{\text{vib}}/T)$ in powers of T_{vib}/T. By arresting this expansion to the linear term we immediately get

$$\mathcal{U}_{\text{vib}} = \frac{1}{2}nRT_{\text{vib}} + nRT \tag{2.16}$$

which is understood to be valid whenever $T > T_{\text{vib}}$.

By combining equations (2.12), (2.14), and (2.16) we get the asymptotic value of the molar heat capacity of the diatomic molecular gas

$$C_V = \frac{3}{2}R + R + R \tag{2.17}$$

to be compared with equation (2.11) and valid for very high temperatures.

2.1.4 Equipartition of energy

The careful analysis of the various energy contributions in a diatomic molecule we have developed in the previous sections paves the way to a very important result which is known as the *equipartition theorem* or *principle of equipartition of energy* which can be formulated in a very simple way: *in equilibrium at temperature T, the mean value of each independent quadratic term in the energy is worth of $k_{\mathrm{B}}T/2$.* It is important to remark that this result holds only for classical systems, where the energy spectrum is a continuum. In view of its importance, we will present both a phenomenological and a formal proof of this theorem, while more fundamental discussions can be found elsewhere [3–6].

2.1.4.1 Phenomenological argument

In a diatomic gas each molecule has seven degrees of freedom, namely: three translational (associated with the drift motion of its centre of mass), two rotational (associated with rigid rotations around the two axes normal to the molecular one), and two vibrational (associated with the oscillation along the molecular axis of each atom around its equilibrium position).

All degrees of freedom are *quadratic either in the particle positions or in the particle momenta* since: the translational ones involve kinetic energy and rotational ones the square of the angular momentum, while a bit more subtle is the case of the vibrational degrees of freedom. Here a more convenient description is provided by observing that the vibrational physics of a diatomic molecule actually corresponds to a one-dimensional oscillator [1, 2] whose energy does contain a kinetic and a potential energy term, the latter being quadratic in the relative atom–atom distance under the harmonic approximation.

We have already proved that in equilibrum condition at temperature T the average kinetic energy of a structureless particle is $\langle E \rangle = 3k_{\mathrm{B}}T/2$, as shown in section 1.5. This results also holds for the translational motion of a rigid and non-rotating diatomic molecule; accordingly, we can draw the conclusion that *each translational degree of freedom on average carries an energy $k_{\mathrm{B}}T/2$.* By generalisation, we attribute the same amount of average energy to each rotational and to each vibrational degree of freedom, simply relying on the fact that they all are quadratic like the translational ones. This is tantamount to stating that on average a diatomic molecule has a total energy as large as $\langle E \rangle = 3k_{\mathrm{B}}T/2 + k_{\mathrm{B}}T + k_{\mathrm{B}}T = 7k_{\mathrm{B}}T/2$, as indeed found experimentally and shown in the high-temperature regime reported in figure 2.1 which we can now interpret by a very effective physical picture. At very low temperatures $T \ll T_{\mathrm{rot}} \ll T_{\mathrm{vib}}$ both the rotational and the vibrational degrees of freedom are 'frozen' (that is, their 'effective temperature' is zero) and by virtue of the equipartition principle do not contribute to the internal energy of the gas; accordingly, only translational energy is at work and the molar heat capacity is the same as the monoatomic gas. At intermediate temperatures $T_{\mathrm{rot}} \ll T \ll T_{\mathrm{vib}}$ only rotational degrees of freedom are activated and the resulting molar heat capacity is now $5R/2$, corresponding to the first two terms of equation (2.17). Eventually, at higher temperatures $T \gg T_{\mathrm{vib}} \gg T_{\mathrm{rot}}$ all

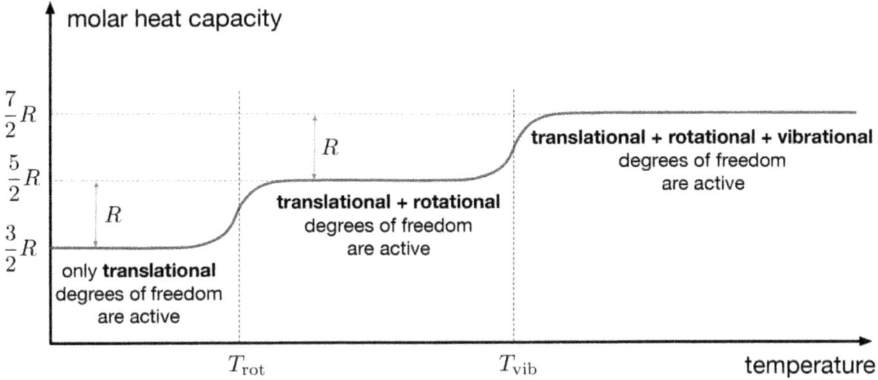

Figure 2.1. Qualitative trend of variation of the molar heat capacity of a diatomic molecular gas against increasing temperature. T_{rot} and T_{vib} are the intrinsic characteristic rotational and vibrational temperature of the molecule, respectively. See text for more details.

molecular degrees of freedom are activated, including the vibrational ones, and the total molar heat capacity is as large as $7R/2$.

2.1.4.2 Formal proof

Let us consider a classical system with M quadratic degrees of freedom which are described by general coordinates ξ_i with $i = 1, 2, \ldots, M$. Following the above phenomenological arguments, we understand that these general coordinates could be either particle positions or particle momenta. To keep the formalism as clean as possible, we will label by $\{\Xi\}$ the full set of such variables.

Let us further assume that the total internal energy $\mathcal{U}$ of the system is additive in the quadratic degrees of freedom and accordingly write

$$\mathcal{U}(\{\Xi\}) = \sum_{i=1}^{M} c_i \xi_i^2 = \sum_{i=1}^{M} \epsilon_i \tag{2.18}$$

where $\epsilon_i = c_i \xi_i^2$ is the energy associated with the ith degree of freedom. We aim at computing its average value $\langle \epsilon_i \rangle$ when the system is in an equilibrium condition at temperature T. We will make use of the compact notation $\beta = 1/k_{\mathrm{B}} T$.

In order to proceed we must generalise the arguments developed in section 1.5 to calculate the average value of a physical observable. Since ξ_i's are continuous variables we replace the sums appearing in equation (1.30) by integrals obtaining

$$\langle \epsilon_i \rangle = \frac{\displaystyle\int \epsilon_i \exp\left[-\beta \mathcal{U}(\{\Xi\})\right] d\xi_1 d\xi_2 \cdots d\xi_M}{\displaystyle\int \exp\left[-\beta \mathcal{U}(\{\Xi\})\right] d\xi_1 d\xi_2 \cdots d\xi_M} \tag{2.19}$$

where the partition function is now written as $\mathcal{Z} = \int \exp[-\beta \mathcal{U}(\{\Xi\})] d\xi_1 d\xi_2 \ldots d\xi_M$ and, by following the arguments developed in section 1.6.3, all intrinsic probabilities are set equal. The calculation of $\langle \epsilon_i \rangle$ straightforwardly proceeds from equation (2.19) by using equation (2.18)

$$
\begin{aligned}
\langle \epsilon_i \rangle &= \frac{\int \epsilon_i \exp[-\beta \sum_{i=1}^{M} \epsilon_i] d\xi_1 d\xi_2 \cdots d\xi_M}{\int \exp[-\beta \sum_{i=1}^{M} \epsilon_i] d\xi_1 d\xi_2 \cdots d\xi_M} \\[2mm]
&= \frac{\int \epsilon_i \exp[-\beta \epsilon_i] d\xi_i \ \int \exp[-\beta \sum_{j(\neq i)=1}^{M} \epsilon_j] d\xi_1 d\xi_2 \ldots d\xi_{i-1} d\xi_{i+1} \cdots d\xi_M}{\int \exp[-\beta \epsilon_i] d\xi_i \ \int \exp[-\beta \sum_{j(\neq i)=1}^{M} \epsilon_j] d\xi_1 d\xi_2 \ldots d\xi_{i-1} d\xi_{i+1} \cdots d\xi_M} \\[2mm]
&= \frac{\int \epsilon_i \exp[-\beta \epsilon_i] d\xi_i}{\int \exp[-\beta c_i] d\xi_i} = \frac{c_i \int \xi_i^2 \exp[-\beta c_i \xi_i^2] d\xi_i}{\int \exp[-\beta c_i \xi_i^2] d\xi_i} = \frac{\frac{c_i}{4}\sqrt{\frac{\pi}{(\beta c_i)^3}}}{\frac{1}{2}\sqrt{\frac{\pi}{\beta c_i}}}
\end{aligned}
\tag{2.20}
$$

where in the last step we made use of the results for the Gaussian integrals reported in appendix A. We eventually come to the result

$$
\langle \epsilon_i \rangle = \frac{1}{2\beta} = \frac{1}{2} k_{\mathrm{B}} T
\tag{2.21}
$$

which represents the formal calculation of the average value of any quadratic term in the energy for a classical system.

As a showcase application of the equipartition theorem we calculate the constant-volume molar heat capacity of a monoatomic crystalline solid. Classically, we can describe such a system as an assembly of N atoms harmonically oscillating around fixed lattice positions [7]. Accordingly, each crystalline atom is a three-dimensional harmonic oscillator counting six quadratic terms in the energy: three kinetic and three potential. According to equation (2.21) the atomic mean energy is $\langle \epsilon \rangle = 3k_{\mathrm{B}} T$, just the same for any particle in the crystal. For a crystal containing as many as $N = n \mathcal{N}_{\mathrm{A}}$ atoms, the resulting internal energy is $\mathcal{U} = 3n \mathcal{N}_{\mathrm{A}} k_{\mathrm{B}} T$ so that $C_{\mathrm{V}} = 3R$, a result known as *Dulong–Petit law* according to which the heat capacity of a solid does not depend on temperature. Experiments only validate this result in the high-remperature regime, while a $C_{\mathrm{V}} = C_{\mathrm{V}}(T)$ dependence at intermediate and low temperatures is observed. However, this is not a failure of the equipartition theorem, but an invalidation of the assumption that the vibrational properties of the crystal can be treated classically even at low temperatures. This issue will be readdressed in section 4.2.2.

2.2 Real gases

We conclude our journey into the thermal properties of gaseous systems by considering the case of a *real gas* where, at variance with the constitutive hypotheses of the ideal model discussed in chapter 1, we recognise that *its elementary constituents undergo mutual interactions*. Since here lies the main qualitative difference between the ideal and real cases, we will treat in detail the paradigmatic case of a monoatomic gas with particle–particle interactions[3]. More specifically, we will assume that the force field acting within the system is given by the superposition of short-ranged, two-body and central[4] interatomic potentials.

Despite these simplifications, the model is good enough to catch the main differences between the equation of state of a real gas and equation (1.55) valid in the ideal case. Of course in real physical systems particles (either atomic or molecular) do have a finite volume: in other words, the volume that a particle can move around is not exactly the volume of the container within which the gas is confined. This 'excluded volume' effect is not treated here (or, equivalently, the atoms—although interacting—are still assumed point-like), but it is treated elsewhere where a more thorough theory is developed [3, 4].

2.2.1 Equation of state

As recognised in section 1.7.2, the key quantity needed to calculate the equation of state is the partition function. For the ideal gas it is provided by equation (1.52) which, when considering interatomic interactions, is clearly missing a potential energy term: equivalently, we can state that equation (1.52) does only include the kinetic energy contribution, the only one actually present in a gas of non-interacting particles. Therefore, in order to write a workable expression of the partition function for a real gas, we should properly generalise this result so as to include the potential energy

$$E_{\text{pot}} = \sum_{\alpha > \beta} V(R_{\alpha\beta}) \tag{2.22}$$

where $\mathbf{R}_\alpha$ (with $\alpha = 1, 2, 3, \ldots, N$) are the positions of the N atoms, $R_{\alpha\beta}$ is the distance between the αth and the βth particles, while $V(R_{\alpha\beta})$ is the two-body interatomic potential for this pair. Another preliminary step is to make use of the *grand partition function* which, for the ideal gas, is defined as

$$\bar{Z}_{\text{ideal}} = \frac{1}{N!} Z_{\text{ideal}}^N = \frac{1}{N!} \left[\frac{V(2\pi m k_{\text{B}} T)^{3/2}}{h^3} \right]^N \tag{2.23}$$

where we make use of equation (1.52) for Z_{ideal}; this is a more convenient partition function to be used to calculate the statistical definition of pressure as

$$P = k_{\text{B}} T \frac{\partial \ln \bar{Z}}{\partial V} \bigg|_T \tag{2.24}$$

which turns out to be perfectly consistent with the previous definition provided in equation (1.54), as simply shown by a direct application to the ideal gas case.

---

[3] Of course, if the real gas is made by molecules, the roto-vibrational degrees of freedom should also be included in the theory. However, for the sake of simplicity we limit our discussion to the case of an atomic gas.
[4] That is, only depending on the actual distance between the two interacting particles.

After the above preliminary considerations, we are now ready to write the *grand partition function $\bar{Z}_{\text{real}}$ of a gas of interacting atoms*. While its detailed calculation falls beyond the scope of this Primer [3, 4], we will proceed in analogy with the ideal gas case by fully exploiting a basic fact which has been so far left unsaid: if the total energy of a system is given by the sum of different contributions, for each of which a specific partition function can be calculated, then *the total partition function is the product of the specific ones*[5]. Now, the total energy of a real interacting gas is the sum of a kinetic energy contribution, to be treated as in the ideal case, and a potential energy contribution. This implies that $\bar{Z}_{\text{ideal}}$ must be corrected by a multiplication of an integral which has the general form as in equation (1.51), but for the twofold fact that (i) it contains E_{pot} and (ii) it implies integration of the space coordinates $\mathbf{R}_\alpha$

$$\bar{Z}_{\text{real}} = \frac{1}{N!}\left[\frac{(2\pi m k_{\text{B}}T)^{3/2}}{h^3}\right]^N \int \exp\left(-E_{\text{pot}}/k_{\text{B}}T\right)d\mathbf{R}_1 d\mathbf{R}_2 \cdots d\mathbf{R}_N \qquad (2.25)$$

where an important detail must be commented: if we proceed by setting $E_{\text{pot}} = 0$, the integral appearing in this equation just provides a V^N contribution. This makes equation (2.25) consistent with equation (2.23) in the case of ideal gas. In other words, equation (2.25) is the correct grand partition function for a real gas, recovering the proper form in the case of missing interatomic interactions.

The challenge is now to calculate the multiple integral appearing in equation (2.25) for the case $E_{\text{pot}} \neq 0$ of interest. This is a lengthy calculation explicitly leading to

$$\int \exp\left(-E_{\text{pot}}/k_{\text{B}}T\right)d\mathbf{R}_1 d\mathbf{R}_2 \cdots d\mathbf{R}_N = V^N\left(1 + \frac{N^2 a}{2V}\right) \qquad (2.26)$$

as shown in appendix D; here we limit ourselves to stressing that the a term is just a constant depending on the interatomic potential $V(R_{\alpha\beta})$. By inserting this expression into equation (2.25) we get the explicit form of the grand partition function for the real gas

$$\bar{Z}_{\text{real}} = \frac{1}{N!}\left[\frac{(2\pi m k_{\text{B}}T)^{3/2}}{h^3}\right]^N V^N\left(1 + \frac{N^2 a}{2V}\right) \qquad (2.27)$$

whose natural logarithm can be easily calculated

$$\begin{aligned}
\ln \bar{Z}_{\text{real}} &= \ln \frac{1}{N!}\left[\frac{(2\pi m k_{\text{B}}T)^{3/2}}{h^3}\right]^N + N \ln V + \ln\left(1 + \frac{N^2 a}{2V}\right) \\
&\simeq \ln \frac{1}{N!}\left[\frac{(2\pi m k_{\text{B}}T)^{3/2}}{h^3}\right]^N + N \ln V + \frac{N^2 a}{2V}
\end{aligned} \qquad (2.28)$$

---

[5] For instance, this result has been implicitly used in the calculation of the molar constant-volume heat capacity of a molecular gas for which $Z_{\text{tot}} = Z_{\text{tr}}Z_{\text{rot}}Z_{\text{vib}}$.

where we approximated $\ln(1 + x) \simeq x$ since[6] $x = N^2a/2V \ll 1$. This is what is actually needed to eventually evaluate the pressure according to its statistical definition provided in equation (2.24)

$$P = k_{\rm B}T\frac{\partial \ln \bar{\mathcal{Z}}_{\rm real}}{\partial V}\bigg|_{T} = \frac{k_{\rm B}TN}{V} - \frac{k_{\rm B}TN^2a}{2V^2} \tag{2.29}$$

which for $N = n\mathcal{N}_{\rm A}$ becomes

$$P = \underbrace{\frac{nRT}{V}}_{\text{ideal gas}} - \underbrace{\frac{n^2RT\mathcal{N}_{\rm A}a}{2V^2}}_{\text{first correction to ideal gas}} \tag{2.30}$$

clearly showing the correction to the ideal case result provided by interatomic interactions. This is the first-order-corrected *equation of state of a real monatomic gas* made by point-like particles undergoing two-body, short-range, and central interactions. More refined expressions are obtained by considering non-zero atomic volumes and/or long-range interactions [3, 4].

2.2.2 The virial expansion

The result reported in equation (2.30) suggests that the additional physical features we added in our definition of real gas (basically, the interatomic interactions) appear in the equation of state as an additive correction to $P = nRT/V$. By a close inspection of equation (2.30), we could argue even more, namely: all terms contain powers of n/V. Moreover, by physical intuition we could further guess that just one corrective term has appeared simply because of the truncations we operated in deriving equation (4.14) (see appendix D). These arguments naturally lead us to state that *in a real gas the pressure can be cast in the form of a series of negative powers of the molar volume V/n*

$$P = \xi_1\left(\frac{n}{V}\right) + \xi_2\left(\frac{n}{V}\right)^2 + \xi_3\left(\frac{n}{V}\right)^3 + \cdots \tag{2.31}$$

which is hereafter referred to as the *virial expansion for the equation of state of a real gas*. It is understood that $\xi_1 = RT$ in any case, while in order to get the actual values of the other ξ_i ($i = 2, 3, 4, \ldots$) *virial coefficients* we need to develop a full theory for the real gas[7]. For instance, the level of erudition we developed in section 2.2.1 allows us to define

$$\xi_2 = -\frac{RT\mathcal{N}_{\rm A}a}{2} \tag{2.32}$$

as the second virial coefficient.

---

[6] In normal pressure and temperature conditions we typically have $x = \mathcal{O}(10^{-4})$.
[7] Obviously, for an ideal gas all virial coefficients are zero, except for the first one.

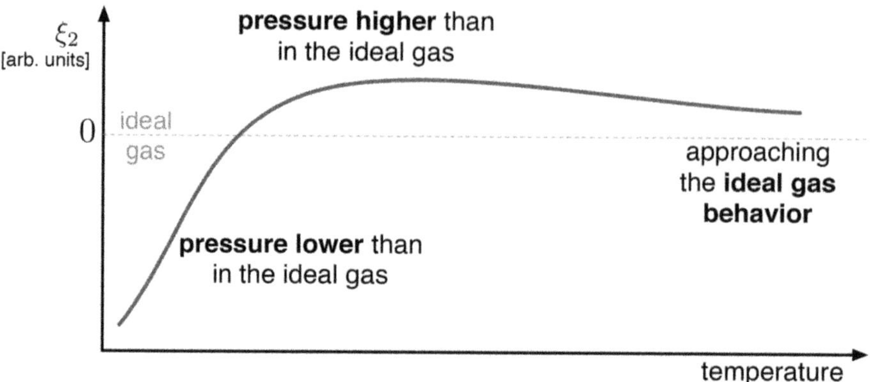

Figure 2.2. Qualitative trend of variation of the second virial coefficient ξ_2 against temperature. For an ideal gas we have $\xi_2 = 0$. See text for discussion.

Similarly to ξ_2, all virial coefficients depend on the temperature, as well as on the force field acting among the particles of the gas (in the case of the second coefficient, this dependence is introduced by the a term, as shown in appendix D). Experimentally, they can be obtained by measuring the pressure of the gas at different temperatures; in this way, the accuracy of the model adopted to describe interatomic forces can be directly verified. In figure 2.2 it is reported the variation upon temperature of the second virial coefficient ξ_2 for noble gases whose interactions are described by means of the so-called Lennard-Jones two-body potential

$$V_{\mathrm{LJ}}(R) = V_0 \left[\left(\frac{\sigma}{R} \right)^{12} - 2\left(\frac{\sigma}{R} \right)^{6} \right] \tag{2.33}$$

where R is the particle–particle distance, while V_0 and σ are two adjustable parameters, respectively defining the depth and the position of the minimum of such a pair potential.

The observed trend is easily understood: at low temperatures where $k_{\mathrm{B}}T \ll V_0$ the particles are most likely found at mutual distances corresponding to minimum interaction energy; here forces are attractive so that the overall pressure of the real gas is smaller than in the ideal case: therefore, $\xi_2 < 0$ (see equation (2.31)). On the other hand, at high enough temperatures $k_{\mathrm{B}}T \gg V_0$ thermal motion brings particles very close together, so that they experience the strongly repulsive part of the Lennard-Jones potential; this, in turn, makes the real gas pressure exceed that of the ideal system: therefore, $\xi_2 > 0$. Eventually, at really very high temperatures, the kinetic energy of the particles is so large that they can occasionally approach very close, overcoming the repulsive barrier. This situation mimics what happens in an ideal gas where particles can approach up to infinitesimally short distances: accordingly, $\xi_2 \to 0$. Experimental measurement of the second virial coefficient provides convincing evidence that the Lennard-Jones potential is very accurate in modelling noble gases over a wide range of temperatures [8].

2.3 Thermal properties under the action of an external field

We conclude this chapter by investigating *how a gas behaves under the action of an external field*. We will consider two paradigmatic cases, namely: (i) an atomic gas whose particles have a magnetic moment (this corresponds to the case of any atom in a state other than $^1S_0$, where we have used the standard spectroscopic notation [1, 2] for the electronic configuration) under the action of a magnetic field; and (ii) a molecular gas whose particles have a permanent electric dipole moment (this is the case of polar diatomic molecules) under the action of an electric field. In both cases we assume there are *no interactions among atomic magnetic moments or molecular electric ones*, thus mimicking the physical situation of a dilute gas. The two situations we are about to consider will allow us to understand the basics of paramagnetism and dielectric polarization, respectively. With this important difference: while the magnetic moments undergo spatial quantisation, their electrical counterparts are not subject to this constraint.

2.3.1 Paramagnetism

Let us consider a monoatomic gas made by particles with magnetic moment $\mathbf{M}$ whose quantum mechanical expression [1, 2] is

$$\mathbf{M} = -\frac{\mu_B}{\hbar}\, g\, \mathbf{J} \tag{2.34}$$

where $\mathbf{J}$ is the total (orbital + spin) angular momentum of the atom, $\mu_B = e\hbar/2m_e$ is the Bohr magneton (expressed in terms of the charge e and of the mass m_e of the electron) and

$$g = 1 + \frac{j(j+1) - l(l+1) + s(s+1)}{2j(j+1)} \tag{2.35}$$

is the Landé g-factor (expressed in terms of the quantum numbers l, s and j, respectively, associated with the orbital, spin, and total angular momentum of the atom).

Under the action of a uniform, homogeneous, and constant magnetic field $\mathbf{B} = (0, 0, B)$ any atomic magnetic moment tends to align along such a field. This process generates a net magnetisation or, equivalently, *a magnetic moment per unit volume $\mathcal{M}$ along the field is generated within the system*. The orientations of the atomic magnetic moments are, however, constrained by a spatial quantisation rule: only $(2j + 1)$ possible orientations are in fact possible with respect to the direction of the external magnetic field, corresponding to the allowed values of the magnetic quantum number $m_j = -j, -j + 1, \ldots, j - 1, j$. Therefore, the resulting magnetisation is given by

$$\mathcal{M} = \frac{N}{V}\langle M \rangle \tag{2.36}$$

where N is the number of particles in the gas, V the occupied volume, and $\langle M \rangle$ is the average value of the atomic magnetic moment calculated over all the allowed values $M_j = -\mu_B g m_j$ selected by the spatial quantisation rule. If the gas is in equilibrium at temperature T we calculate $\langle M \rangle$ according to equation (1.30)

$$\langle M \rangle = \frac{\sum_{m_j=-j}^{+j} M_j \exp(-E_{\text{mag}}/k_B T)}{\sum_{m_j=-j}^{+j} \exp(-E_{\text{mag}}/k_B T)} \tag{2.37}$$

where the magnetic partition function is

$$\mathcal{Z} = \sum_{m_j=-j}^{+j} \exp(-E_{\text{mag}}/k_B T) \tag{2.38}$$

and

$$E_{\text{mag}} = -\mathbf{M} \cdot \mathbf{B} = \mu_B g m_j B \tag{2.39}$$

represents the magnetic interaction energy between the field and the atomic magnetic moment. The calculation of equation (2.37) proceeds straightforwardly

$$
\begin{aligned}
\langle M \rangle &= \frac{\sum_{m_j=-j}^{+j}\left(-\mu_B g m_j\right)\exp\left(-\mu_B g m_j B/k_B T\right)}{\sum_{m_j=-j}^{+j} \exp\left(-\mu_B g m_j B/k_B T\right)} \\[2mm]
&= \frac{\sum_{m_j=-j}^{+j}\left(-\mu_B g m_j\right)\left(1 - \mu_B g m_j B/k_B T + \cdots\right)}{\sum_{m_j=-j}^{+j}\left(1 - \mu_B g m_j B/k_B T + \cdots\right)} \\[2mm]
&\simeq \frac{-\mu_B g\left(\sum_{m_j=-j}^{+j} m_j\right) + \dfrac{\mu_B^2 g^2 B}{k_B T}\left(\sum_{m_j=-j}^{+j} m_j^2\right)}{\left(2j + 1\right) - \dfrac{\mu_B g B}{k_B T}\left(\sum_{m_j=-j}^{+j} m_j\right)}
\end{aligned}
\tag{2.40}
$$

and we eventually get

$$\langle M \rangle = \frac{\mu_B^2 g^2 B}{(2j + 1)k_B T}\left(\sum_{m_j=-j}^{+j} m_j^2\right) \tag{2.41}$$

by taking into account that $\sum_{m_j=-j}^{+j} m_j = 0$. The term in square parenthesis is easily calculated

$$\sum_{m_j=-j}^{+j} m_j^2 = (2j + 1)(j + 1)\frac{j}{3} \tag{2.42}$$

which eventually leads to

$$\langle M \rangle = \frac{j(j + 1)(\mu_B g)^2}{3 k_B T} B \tag{2.43}$$

namely the average value of the atomic magnetic moment along the external magnetic field we were looking for. It is worth noticing that $\langle M \rangle$ is linear in the ratio B/T: this is a consequence of the implicit assumption of weak magnetic field we did in developing equation (2.40) where the exponential term was expanded in powers of B/T and arrested just to the linear term. Interestingly enough, the same result is obtained by assuming very low temperatures. This point will be critically readdressed below.

The *paramagnetic magnetisation* is accordingly predicted to be

$$\mathcal{M} = \frac{N}{V} \frac{j(j+1)(\mu_B g)^2}{3 k_B T} B \tag{2.44}$$

and the corresponding *Curie paramagnetic susceptibility* χ_C is

$$\chi_C = \frac{\mathcal{M}}{B} = \frac{N}{V} \frac{j(j+1)(\mu_B g)^2}{3 k_B T} = \frac{N M^2}{3 k_B T} \tag{2.45}$$

where we made use of equation (2.34) to calculate the square magnetic moment: $M^2 = \mu_B^2 g^2 J^2 / \hbar^2 = \mu_B^2 g^2 j(j+1)$. Equation (2.45) proves that *in the regime of weak field and/or low temperature, the magnetic susceptibility is independent of the applied magnetic field.*

The situation is qualitatively different if the ratio B/T is so large that the power expansion performed in equation (2.40) is no longer justified. We treat this case in the limit of *very large values of the quantum number j*: accordingly, the sums appearing in equation (2.40) can be replaced by integrals. The explicit calculation [3, 4] leads to a more general expression for the average atomic magnetic moment

$$\langle M \rangle = j \mu_B g L(x) \tag{2.46}$$

where $x = j \mu_B g B / k_B T$ and

$$L(x) = \coth(x) - \frac{1}{x} \tag{2.47}$$

is known as Langevin function; its plot is reported in figure 2.3 showing that the average atomic magnetic moment is proportional to the orientational magnetic field for small B values, while it saturates to the asymptotic value $\langle M \rangle \rightarrow J \mu_B g$ in the limit of strong field: this saturation corresponds to a perfect alignment of all atomic moments along the direction of the external field. In this condition the Curie susceptibility is given by

$$\chi_C = \frac{N}{V} \frac{(j \mu_B g)^2}{B} \tag{2.48}$$

Usually the field is not that strong and, therefore, just the range of small x values is explored; the Taylor expansion of the Langevin function[8] in the limit $x \rightarrow 0$ is $L(x) \simeq x/3$ so that

---

[8] Just remember that $\coth(x) = 1/x + x/3 + \cdots$ for any $x < 1$.

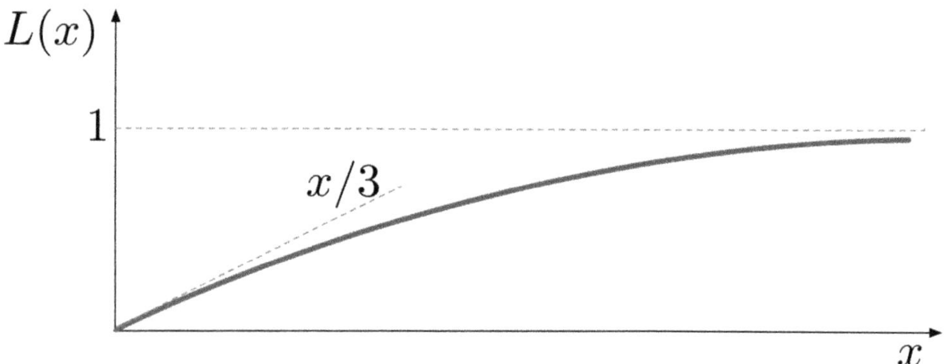

Figure 2.3. Plot of the Langevin function versus the dimensionless variable $x = j\mu_{\mathrm{B}}gB/k_{\mathrm{B}}T$.

$$\langle M \rangle = (j\mu_{\mathrm{B}}g)^2 \frac{B}{3k_{\mathrm{B}}T} \tag{2.49}$$

which is basically just the same as equation (2.43) since we are considering the case of very large j-values for which $j(j + 1) \simeq j^2$.

The problem we have discussed does not in fact correspond to any real physical case, basically because there are no paramagnetic monoatomic gases[9]. However, the results developed here apply well to the case of paramagnetic salts. They are solid-state systems where isolated magnetic ions are randomly distributed on crystal lattice, separated by non-magnetic ions. In this way they reproduce the situation of non-interacting magnetic moments and the system behaviour under the action of an externally applied field is to a very good extent described by the Curie paramagnetic susceptibility.

2.3.2 Polarisation

Let us now consider a gas of polar molecules in equilibrium at temperature T, placed in a uniform, homogeneous, and constant electric field $\mathbf{E}$ and let $\mathbf{p}$ be the electric dipole of each molecule. We will study the case where the electric field is strong enough to orient the molecular dipoles, but not to alter their roto-vibrational or translational state.

In order to calculate the partition function of the system we preliminarily observe that the energy E_{elec} of a dipole under the action of an electric field is

$$E_{\mathrm{elec}} = -\mathbf{p} \cdot \mathbf{E} = -pE \cos \theta \tag{2.50}$$

where θ is the angle between the field and the molecular moment. This angle can have any value in the range $0 \leqslant \theta \leqslant \pi$ (that is, any molecule is free to follow the orienting action of the external field) since, at variance with the magnetic case, no space quantisation phenomena here occur. This implies that the energy given in

---

[9] There are instead paramagnetic molecular gases, such as O_2 gas, which, however, require a much more complex theoretical treatment.

equation (2.50) varies continuously with θ and, accordingly, we should use an integral expression for $\mathcal{Z}$ as we did in section 1.7.1 in the case of the ideal monoatomic gas. In this case the role played by the density of energy levels is played by the *number of molecules with their dipole moment oriented within the solid angle* $d\Theta = 2\pi \sin\theta d\theta$ contained within two cones of angles θ and $\theta + d\theta$. In virtue of this, the partition function is written as

$$\mathcal{Z} = \int_0^\pi \exp(pE \cos\theta/k_BT)2\pi \sin\theta d\theta = 4\pi\frac{k_BT}{pE} \sinh\left(\frac{pE}{k_BT}\right) \tag{2.51}$$

which allows us to evaluate the mean value of the molecular dipole as

$$\langle p \rangle = \frac{1}{\mathcal{Z}} \int_0^\pi p \cos\theta \exp(pE \cos\theta/k_BT)2\pi \sin\theta d\theta \tag{2.52}$$

where of course the physical observable to average is $p \cos\theta$, namely the dipole component along the direction of the field. By calculating the integral and by inserting into this equation the above expression for the partition function we eventually get

$$\langle p \rangle = pL(x) \tag{2.53}$$

with $x = pE/k_BT$; once again, the central role is played by the Langevin function.

Large values of the electric field or, equivalently, very low temperatures correspond to the limit $x \to +\infty$ and, as shown in figure 2.3, this corresponds to $L(x) \to 1$, that is, $\langle p \rangle = p$: all molecular dipoles are parallel and oriented along the direction of the electric field. The polarisation (that is, its total electric dipole moment per unit volume) of the gas reaches its maximum value $\mathcal{P} = Np/V$. On the other hand, whenever E/T is small (a situation corresponding to a weak field and/or high temperature) we instead obtain $\langle p \rangle = p^2 E/3k_BT$ and the resulting polarisation is $\mathcal{P} = Np^2E/3Vk_BT$. This expression is very useful, and in fact extensively adopted, in the calculation of the electrical polarizability of dielectric media.

References

[1] Demtröder W 2010 *Atoms, Molecules and Photons* (Berlin: Springer)
[2] Colombo L 2019 *Atomic and Molecular Physics: A Primer* (Bristol: IOP Publishing)
[3] Kennett M P 2021 *Essential Statistical Mechanics* (Cambridge: Cambridge University Press)
[4] Reif F 1987 *Fundamentals of Statistical and Thermal Physics* (New York: McGraw-Hill)
[5] Huang K 1987 *Statistical Mechanics* (New York: Wiley)
[6] Balescu R 1975 *Equilibrium and Non-equilibrium Statistical Mechanics* (New York: Wiley)
[7] Colombo L 2021 *Solid State Physics: A Primer* (Bristol: IOP Publishing)
[8] Hirschfelder J O, Curtiss C F and Bird R B 1954 *Molecular Theory of Gases and Liquids* (New York: Wiley)

Part II

Quantum statistical physics

IOP Publishing

Statistical Physics of Condensed Matter Systems
A primer
Luciano Colombo

Chapter 3

The statistical description of a quantum system

***Syllabus**—We introduce the basic concepts for the statistical description of a system made by a large number of particles obeying quantum mechanics. We start by considering the symmetric or antisymmetric character of the total wavefunction describing a set of identical and indistinguishable quantum particles. This will allow us to make the fundamental separation between fermions and bosons. Next, we thoroughly derive the Fermi–Dirac and the Bose–Einstein statistics, respectively, valid for particles with half-odd integer spin and integer spin. They describe the equilibrium state of a quantum system. The chapter ends with a unified derivation of both quantum statistics based on the grand canonical ensemble.*

3.1 Basic concepts

3.1.1 Symmetric and antisymmetric wavefunctions

Two particles are identical provided that they have entirely the same characteristics (i.e. same intrinsic physical properties like, for instance, the mass, the charge, the spin, and so on). Identical particles can be distinguished, if treated according to classical mechanics, by taking into consideration their positions. On the other hand, according to quantum mechanics we cannot assign a well defined position to any particle and this prevents distinguishing between them: in short, *in quantum mechanics identical particles are inherently indistinguishable.*

Let us then consider the Hamiltonian operator $\hat{H}$ describing a system of N *identical and indistinguishable particles*: it is easy to understand that it is invariant upon interchanging any two particles. However, the same operation could differently affect its eigenfunction $\Phi(\mathbf{r}_1, \mathbf{r}_2, \ldots, \mathbf{r}_N)$ describing the state of the system. In stating that $\Phi(\mathbf{r}_1, \mathbf{r}_2, \ldots, \mathbf{r}_N)$ is an eigenfunction of $\hat{H}$, we understand that

doi:10.1088/978-0-7503-2269-0ch3

$$\hat{H}\Phi(\mathbf{r}_1, \mathbf{r}_2, \ldots , \mathbf{r}_N) = E_T\Phi(\mathbf{r}_1, \mathbf{r}_2, \ldots , \mathbf{r}_N) \tag{3.1}$$

where E_T is the total energy of the system of particles. This situation is worth further consideration.

Let $\hat{P}_{\mu\leftrightarrow\nu}$ be the *permutation operator* which acts on the eigenfunctions of $\hat{H}$ by swapping the particles labelled by the indices $\mu \leftrightarrow \nu$. Because of the above invariance the two operators $\hat{H}$ and $\hat{P}_{\mu\leftrightarrow\nu}$ commute

$$[\hat{H}, \hat{P}_{\mu\leftrightarrow\nu}] = 0 \tag{3.2}$$

and, therefore, $\Phi(\mathbf{r}_1, \mathbf{r}_2, \ldots , \mathbf{r}_N)$ *is a wavefunction of either $\hat{H}$ and $\hat{P}_{\mu\leftrightarrow\nu}$*. It is easy to prove that

$$\hat{P}_{\mu\leftrightarrow\nu}\Phi(\mathbf{r}_1, \mathbf{r}_2, \ldots , \mathbf{r}_N) = \pm\Phi(\mathbf{r}_1, \mathbf{r}_2, \ldots , \mathbf{r}_N) \tag{3.3}$$

or, equivalently: any wavefunction describing a set of identical and indistinguishable particles is either symmetric (eigenvalue $+1$) or antisymmetric (eigenvalue -1) under particle interchange.

It is very important to remark that *the symmetry property of the total wavefunction reflects a fundamental intrinsic characteristic of the specific set of particles considered* since it cannot be altered by any external action. Identical and indistinguishable particles are said to be *fermions* or *bosons* if their collective quantum states are described by antisymmetric or symmetric wavefunctions, respectively [1–4].

The antisymmetric character of the fermion wavefunction has a very important consequence which we will derive under the assumption that a system of N fermions can be treated within a single-particle approximation [5]; in other words, we will assume that the complete problem given in equation (3.1) can be reduced to N different single-particle problems

$$\hat{h}_i\, \phi_{\alpha_i}(\mathbf{r}_i) = E_i\phi_{\alpha_i}(\mathbf{r}_i) \tag{3.4}$$

where $i = 1, 2, \ldots , N$ labels the fermions, ϕ's are the single-particle state functions, and α's are the corresponding sets of quantum numbers (including the spin quantum number). Of course within this single single-particle approach we have $E_T = \sum_i E_i$.

In order to exploit its antisymmetric character we write the total wavefunction in the form of a Slater determinant

$$\Phi(\mathbf{r}_1, \mathbf{r}_2, \ldots , \mathbf{r}_N) = \frac{1}{\sqrt{N!}} \begin{vmatrix} \varphi_{\alpha_1}(\mathbf{r}_1) & \varphi_{\alpha_1}(\mathbf{r}_2) & \cdots & \varphi_{\alpha_1}(\mathbf{r}_N) \\ \varphi_{\alpha_2}(\mathbf{r}_1) & \varphi_{\alpha_2}(\mathbf{r}_2) & \cdots & \varphi_{\alpha_2}(\mathbf{r}_N) \\ \cdots & \cdots & \cdots & \cdots \\ \varphi_{\alpha_N}(\mathbf{r}_1) & \varphi_{\alpha_N}(\mathbf{r}_2) & \cdots & \varphi_{\alpha_N}(\mathbf{r}_N) \end{vmatrix} \tag{3.5}$$

which naturally embodies the antisymmetric character of the state function, since by swapping two columns the Slater determinant will change sign. In addition, we remark that a determinant vanishes if two rows are equal. This implies that *a system of identical and indistinguishable fermions cannot occupy a many-body state where two single-particle states are equal*. This statement is commonly referred to as *Pauli principle* or, alternatively, exclusion principle; it states that it is impossible to have two fermions with the same set of quantum numbers[1].

The rationalisation of a large number of different experimental investigations leads to the conclusion that electrons, protons, and neutrons are fermions, while photons and phonons are bosons [6]. The fermion or boson character of an elementary particle is ultimately determined by its spin [1, 3, 4, 7, 8]: fermions have half-odd integer spin $\hbar/2$, $3\hbar/2$,... while bosons have integer spin 0, $\hbar$,....

In quantum statistics the ultimate goal is to find the equilibrium distribution for a system of identical and indistinguishable fermions (subjected to the exclusion principle) or bosons (unaffected by any exclusion rule). To this aim we will adopt the single-particle picture [9] and assume a discrete energy spectrum, where single-particle states can possibly be degenerate. Furthermore, our approach will be inspired to the classical theory developed in chapter 1: first, we will evaluate the number of different ways to distribute the fermions or bosons among the available energy levels; next, their most probable partition will be determined as that with the maximum number of different realisations; eventually, the equilibrium distribution will be identified as the most probable one. We will separately treat the case of fermions and bosons since they are inherently different.

3.1.2 Intrinsic probability of a quantum state

In quantum mechanics the very concept of intrinsic probability p_i for the ith single-particle state introduced in section 1.2 must be critically readdressed: p_i must be now intended as *the number of different single-particle quantum states with the same energy*. In other words, p_i is given by the state degeneracy.

In order to elucidate this concept we consider the simple case of an electron in a central potential field: quantum mechanics dictates that each energy level is $(2l + 1)$-fold degenerate, where l is the orbital quantum number [5, 10]; this kind of degeneracy, named 'necessary', is associated with the $(2l + 1)$ different values that the magnetic quantum number m_l can assume. If the central potential is, in particular, Coulomb-like then the electron energy also does not depend on $l = 0, 1, 2,..., n$ where n is the principal quantum number; this kind of degeneracy is named 'accidental'. We conclude that an energy level of an electron undergoing a pure Coulomb potential (like in the hydrogen atom) with principal quantum number n has a total degree of degeneracy as large as $\sum_{l=0}^{n}(2l + 1) = n^2$ which in fact corresponds to its intrinsic probability. If the spin degree of freedom is added

---

[1] The Pauli principles is also formulated in a somewhat weaker form: if we exclude the spin quantum number from the set α_i, then according to Pauli principle we can accommodate up to two fermions with paired spin (that is, the two spins must be antiparallel) on each single-particle state.

(and no magnetic interactions are considered), then the degeneracy further increases up to $2n^2$.

We remark that although p_i states may have the same energy, they really correspond to different physical states since they are described by different wavefunctions and, therefore, other physical observables may have different values.

Finally, in the specific case of fermions subjected to the Pauli exclusion principle, p_i represents the maximum number of particles that can be placed on such an energy state. This implies that, when developing our statistical arguments, the number n_i of fermions that can be accommodated on the ith level is restricted to $n_i \leqslant p_i$.

3.2 Fermi–Dirac statistics

3.2.1 The equilibrium partition

Let us consider a large system of N fermions, distributed on discrete E_1, E_2, E_3,... single-particle energy levels with p_1, p_2, p_3,... degeneracy, respectively. In order to populate the ith energy level with $n_i \leqslant p_i$ fermions we proceed step by step, at first considering the particles as distinguishable: this is just a simplifying trick introduced to make our reasoning simple, but it will be soon removed. We start by placing the first particle on a randomly selected state among the p_i available ones; there are exactly p_i different options for doing that. Next, the second particle is placed on any one of the remaining $(p_i - 1)$ empty states; this time we have $(p_i - 1)$ different possible ways to do that. By proceeding in this way we get

$$p_i(p_i - 1)(p_i - 2)\cdots(p_i - n_i + 1) = \frac{p_i!}{(p_i - n_i)!} \tag{3.6}$$

as the total number of different ways to place n_i distinguishable fermions on this p_i-fold degenerate state. If we now recover the notion that the quantum identical particles are actually indistinguishable, we must divide this number by the number $n_i!$ of pair permutations within the assigned set of n_i particles. In conclusion, the *total number Ω of different ways to place n_1 indistinguishable fermions on the energy level E_1 (which is p_1-fold degenerate), n_2 indistinguishable fermions on the energy level E_2 (which is p_2-fold degenerate), and so on is*

$$\Omega = \prod_i \frac{p_i!}{n_i!(p_i - n_i)!} \tag{3.7}$$

which, as in classical statistical physics, corresponds to a probability

$$P = \xi\Omega \tag{3.8}$$

of realising the assigned partition, where ξ is just a convenience proportionality factor not playing any major role in the theory we are developing.

Even for quantum systems *the equilibrium state is identified with the maximum probability partition* and, therefore, we are now faced to the problem of finding the maximum of P or, equivalently, the maximum of its natural logarithm $\ln P$. As the first step we calculate

$$\ln P = \ln \xi + \ln \left[\prod_i \frac{p_i!}{n_i!(p_i - n_i)!} \right]$$

$$= \ln \xi + \sum_i \ln \frac{p_i!}{n_i!(p_i - n_i)!} \tag{3.9}$$

$$= \ln \xi + \sum_i \left[p_i \ln p_i - n_i \ln n_i - (p_i - n_i)\ln(p_i - n_i) \right]$$

where we used Stirling formula (see appendix A) and from this we eventually obtain

$$-d \ln P = \sum_i \left[\ln n_i - \ln(p_i - n_i) \right] dn_i \tag{3.10}$$

which is the fermion counterpart of equation (1.8). The maximum probability problem, corresponding to setting $-d \ln P = 0$, must be solved under the twofold constrain that the total number of fermions is conserved $\sum_i n_i = N$ and the system is isolated (total energy conserved) $\sum_i n_i E_i = $ constant. According to the Lagrange method (see appendix A), we can write

$$\sum_i \left[\ln n_i - \ln(p_i - n_i) + \alpha + \beta E_i \right] dn_i = 0 \tag{3.11}$$

where α and β are the Lagrange multipliers. The maximum probability is then found by imposing

$$\ln n_i - \ln(p_i - n_i) + \alpha + \beta E_i = 0 \tag{3.12}$$

which leads to

$$n_i = \frac{p_i}{\exp(\alpha + \beta E_i) + 1} \tag{3.13}$$

defining *the most probable (equilibrium) partition for a system of fermions.*

The Lagrange multiplier β plays the same role as in the Boltzmann distribution law and therefore we set again $\beta = 1/k_B T$, where T is the equilibrium temperature of the fermion system. On the other hand, the Lagrange multiplier α must be calculated by imposing that

$$\sum_i \frac{p_i}{\exp(\alpha + E_i/k_B T) + 1} = N \tag{3.14}$$

which clearly suggests that for the sum to have always the same value N we must necessarily have $\alpha = \alpha(T)$. By convention we let

$$\exp(\alpha) = \exp(-\mu_c/k_B T) \tag{3.15}$$

where the quantity μ_c is known as *the chemical potential* and it is itself a function of temperature, as extensively discussed in the next section. In summary, we write

$$n_i = \frac{p_i}{\exp[(E_i - \mu_c)/k_B T] + 1} \tag{3.16}$$

and we refer to this result of paramount importance as *the Fermi–Dirac distribution law* for fermions.

It is customary to define the *the Fermi–Dirac occupation number* $f(E_i, T)$ for the E_i energy level as

$$\frac{n_i}{p_i} = f(E_i, T) = \frac{1}{\exp[(E_i - \mu_c)/k_B T] + 1} \tag{3.17}$$

which by definition is such that $0 \leqslant f(E_i, T) \leqslant 1$ for any possible T value.

We finally remark that, while the derivation of equation (3.16) has been obtained following formal arguments based on the statistical definition of equilibrium (see chapter 1), a more phenomenological and intuitive argument can in fact be developed, leading to the same result. In appendix E we elaborate this model by considering a free electron gas, namely the prototypical fermion system.

3.2.2 The chemical potential

We have seen that the chemical potential μ_c appearing in equation (3.16) is linked to the Lagrange multiplier introduced to enforce the conservation of the total number of particles; accordingly, μ_c *is understood as the thermodynamical parameter ruling over the chemical equilibrium condition* (see appendix C): if two fermion reservoirs are put into contact, then some particles will flow from one to the other until the chemical potentials of the two systems are the same. Its physical meaning is so clarified by thermodynamics: μ_c *provides the work needed to add one particle to the system at constant temperature*, or equivalently

$$\mu_c(T) = \mathcal{F}(T, V, N + 1) - \mathcal{F}(T, V, N) \tag{3.18}$$

where $\mathcal{F}(T, V, N)$ is the Helmholtz free energy of the system of N fermions (see appendix C).

The discussion so far developed clearly indicates that both the Fermi–Dirac occupation number and the chemical potential depend on temperature. It is easy to see from equation (3.17) that

$$\lim_{E \to -\infty} f(E, T) = 1 \qquad \text{and} \qquad \lim_{E \to +\infty} f(E, T) = 0 \tag{3.19}$$

at any non-zero temperature; the width of the energy range over which $f(E, T)$ is non-constant is of the order of $\sim k_B T$. Furthermore, by calculating the derivative

$$\frac{\partial f(E, T)}{\partial E} = -\frac{1}{k_B T} \frac{\exp\left[(E - \mu_c)/k_B T\right]}{\left\{1 + \exp[(E - \mu_c)/k_B T]\right\}^2} \tag{3.20}$$

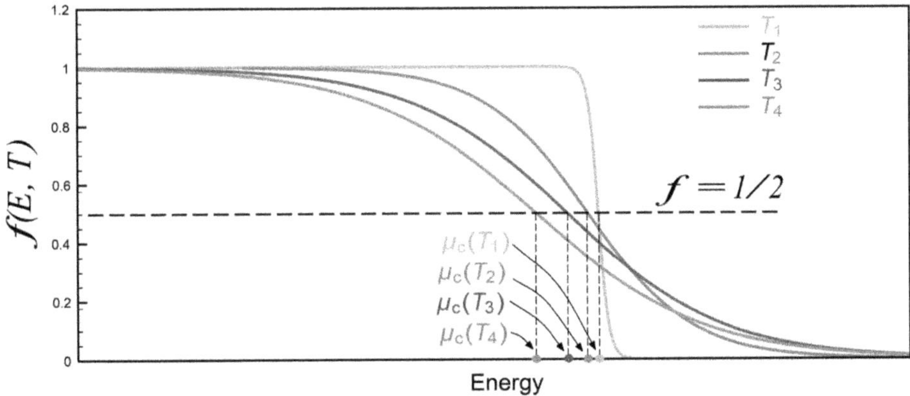

Figure 3.1. The Fermi–Dirac occupation number $f(E, T)$ calculated at four different temperatures $T_1 < T_2 < T_3 < T_4$. The 50% occupation probability (corresponding to $f = 1/2$) is shown by a black dashed line: its intercepts with the curves $f(E, T)$ define the positions of the chemical potential $\mu_c(T)$ at the different temperatures. Reproduced from [9]. © IOP Publishing Ltd. All rights reserved.

we get the maximum value of the slope of the Fermi–Dirac occupation number

$$\left. \frac{\partial f(E, T)}{\partial E} \right|_{E=\mu_c} = -\frac{1}{4k_B T} \tag{3.21}$$

proving that as the temperature decreases, $f(E, T)$ gets closer and closer to a step-like function centred at energy $E = \mu_c$. In particular, at zero temperature we have

$$f(E, 0) = \begin{cases} 1 & \text{for } E < \mu_c \\ 1/2 & \text{for } E = \mu_c \\ 0 & \text{for } E > \mu_c \end{cases} \tag{3.22}$$

We can generalise this result by observing that at any finite temperature it is always found $f(\mu_c, T) = 1/2$: in other words, *the chemical potential corresponds to that energy value which selects the fermion state with 50% occupation probability*. Since $f(E, T)$ is smoothed by increasing the temperature, the chemical potential is correspondingly shifted, as graphically proved in figure 3.1.

We conclude by stressing that the specific temperature-dependence of $\mu_c(T)$ must be elaborated separately for any specific fermion system. In the next chapter we will do that for the electron gas. However, we anticipate a nomenclature widely used: the zero-temperature value of the chemical potential

$$E_F = \lim_{T \to 0} \mu_c(T) \tag{3.23}$$

is referred to as the *Fermi energy* E_F of the system. At zero temperature all energy states $E \leqslant E_F$ are fully occupied (that is, $n_i = p_i$ since their occupation number is 1), while all the states with energy larger than the Fermi energy are empty.

3.3 Bose–Einstein statistics

Let us now move to consider a large system of N bosons, distributed on discrete E_1, E_2, E_3,... single-particle energy levels with p_1, p_2, p_3,... degeneracy, respectively. In this case there is no restriction in the number of particles we can place on each level, since bosons are not subjected to an exclusion principle. Accordingly, the number of different arrangements of n_i bosons on the p_i states with the same energy E_i is given by the number of different ways n_i identical objects can be placed into p_i boxes, with no limit to the number of objects in any single box. This exercise is represented in figure 3.2 in two cases, $p_i = 3$ and $n_i = 2$ (left) or $p_i = 2$ and $n_i = 3$ (right), respectively, corresponding to the case in which the total number of objects is smaller or larger than the total number of boxes. The general formula providing this number is

$$\frac{(n_i + p_i - 1)!}{n_i!(p_i - 1)!} \tag{3.24}$$

so that *the total number Ω of different ways to place n_1 indistinguishable bosons on the energy level E_1 (which is p_1-fold degenerate), n_2 indistinguishable bosons on the energy level E_2 (which is p_2-fold degenerate), and so on* is given by

$$\Omega = \prod_i \frac{(n_i + p_i - 1)!}{n_i!(p_i - 1)!} \tag{3.25}$$

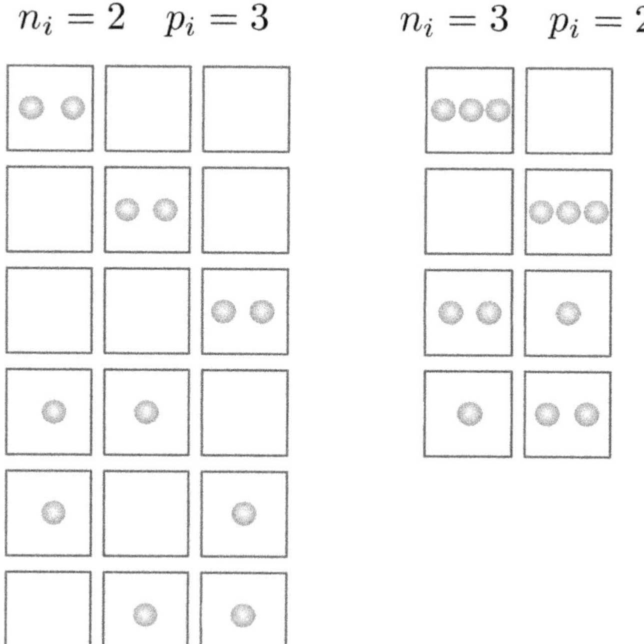

Figure 3.2. Graphical calculation of the number of different ways $n_i = 2$ (left) or $n_i = 3$ (right) identical objects can be placed into $p_i = 3$ (left) or $p_i = 2$ (right) boxes, with no limit to the number of objects in any single box.

defining the probability P for this partition according, as usual, to equation (3.8). Once again we proceed by calculating the natural logarithm of such a probability

$$\ln P = \ln \xi + \ln \left[\prod_i \frac{(n_i + p_i - 1)!}{n_i!(p_i - 1)!} \right] = \ln \xi + \sum_i \ln \frac{(n_i + p_i - 1)!}{n_i!(p_i - 1)!}$$
$$= \ln \xi + \sum_i [(n_i + p_i - 1)\ln(n_i + p_i - 1) - n_i \ln n_i - (p_i - 1)\ln(p_i - 1)] \tag{3.26}$$

where we used Stirling formula (see appendix A). The *equilibrium partition* is obtained by imposing

$$-d \ln P = \sum_i \left[\ln n_i - \ln(n_i + p_i - 1) \right] dn_i = 0 \tag{3.27}$$

which is the boson counterpart of equation (1.8). Next, the $-d \ln P$ quantity is maximised under the constraints of conserved total energy and conserved number of particles. Introducing two suitable Lagrange multipliers for such constraints eventually leads to

$$\ln n_i - \ln(n_i + p_i - 1) + \alpha + \beta E_i \simeq \ln n_i - \ln(n_i + p_i) + \alpha + \beta E_i = 0 \tag{3.28}$$

where we have profited from the fact that $n_i + p_i \gg 1$. We easily get the result

$$n_i = \frac{p_i}{\exp(\alpha + \beta E_i) - 1} \tag{3.29}$$

defining the most probable (equilibrium) partition for a system of bosons or, equivalently, the *Bose–Einstein distribution law*. As in the previous Boltzmann and Fermi–Dirac cases, for the β Lagrange multiplier is set $\beta = 1/k_B T$, while at variance with the fermion statistics the multiplier α has been left undetermined for reasons that are explained just below.

Following the same path as before, we introduce *the Bose–Einstein occupation number $n(E_i, T)$* for the E_i energy level as

$$\frac{n_i}{p_i} = n(E_i, T) = \frac{1}{\exp(\alpha)\exp(E_i/k_B T) - 1} \tag{3.30}$$

where in this case the term $\exp(\alpha)$ must be handled with some care. In fact, quantum systems obeying the Bose–Einstein statistics may or may not have conserved number. In the latter case, there is no need for the Lagrange multiplier α which, accordingly, simply disappears from the theory, as extensively discussed in the case of the phonon gas treated in the next chapter. On the other hand, if the number of bosons is conserved, then α must be explicitly taken into account and calculated; in any case, we remark that since n_i cannot be a negative number, we always have $\alpha > 0$.

3.4 Quantum statistics revisited

In deriving the Fermi–Dirac and Bose–Einstein statistics we have followed the same path previously adopted for classical systems, namely we have looked for the most probable partition providing the equilibrium condition. In doing that we basically counted the total number of different ways to accommodate n_i fermions or bosons on the p_i-fold degenerate energy level E_i; in the first case the Pauli principle was invoked, while no restriction in the number of particles we can place on each level was imposed for bosons.

This way of proceeding is very appealing since it enlightens the most fundamental features of the statistical approach, which are similar for classical and quantum systems (fermion as well as boson). It is undeniable, however, that counting the different ways of realising a given partition is a very technical task and, in some respects, it ends up tarnishing the physical sense. For this reason, in this section we provide an alternative derivation of quantum statistics, based on a thermodynamical approach which should result physically much more transparent. We outline at first the basics of the grand canonical formalism and then heuristically extend it to a gas of fermions and to a gas of bosons by an intuitive guess. A more rigorous treatment can be found elsewhere [11, 12].

3.4.1 The grand canonical ensemble

Let us consider a mono-component quantum system which can exchange energy and particles with a reservoir at constant volume and temperature; according to equation (C.4), the entropy reduction of the reservoir upon a transfer ΔU of energy and ΔN of particles from the reservoir to the system is

$$\Delta S^{\text{res}} = -\frac{1}{T}(\Delta U - \mu_{\text{c}}\Delta N) \tag{3.31}$$

where μ_{c} is the chemical potential of the particles forming the system. Because of this variation, the resulting total entropy of the reservoir is now changed to the value

$$S^{\text{res}} = S_0^{\text{res}} + \Delta S^{\text{res}} \tag{3.32}$$

with respect to the entropy of its initial state S_0^{res}. By using the Boltzmann definition of entropy $S = k_B \ln \Omega$ and recalling the proportionality $P \sim \Omega$ between the probability P of finding a system in a particular state and the number Ω of its corresponding different partitions, we can state that the reservoir is found with relative probability

$$P^{\text{res}} \sim \exp\left[\frac{S_0^{\text{res}}}{k_B} - \frac{(\Delta U - \mu_{\text{c}}\Delta N)}{k_B T}\right] \tag{3.33}$$

in the final state resulting from the above energy and particle transfer to the system.

We now move to consider the system which, after the exchanges with the reservoir, occupies a state s with energy $E_s = \sum_i n_i E_i$ and $N_s = \sum_i n_i$ particles[2]. Therefore, by analogy with equation (3.33), we can write the relative probability of finding the system in such a state s as

$$P_s \sim \exp\left[-\frac{(E_s - \mu_c N_s)}{k_B T}\right] \tag{3.34}$$

while the *absolute probability* is obtained by normalising this expression with respect to all possible particle numbers $N \in [0, +\infty[$ and all possible corresponding states s_N

$$P_s = \frac{1}{\mathscr{Z}_{GC}} \exp\left[-\frac{(E_s - \mu_c N_s)}{k_B T}\right] \tag{3.35}$$

where we have introduced the *grand canonical partition function*

$$\mathscr{Z}_{GC} = \sum_{N=0}^{+\infty} \sum_{s_N} \exp\left\{-\frac{[E_{s_N} - \mu_c N]}{k_B T}\right\} \tag{3.36}$$

and it is understood that, once the number N of particles in the first sum is fixed, the second sum runs over all possible system states characterised by that particle number.

3.4.2 Fermi–Dirac and Bose–Einstein distributions

The grand canonical partition function has been introduced by considering a thermodynamical system interacting with a reservoir of energy and particles. The same conceptual framework can be in fact used to describe a confined gas of quantum particles, regardless of whether fermions or bosons. Elementary quantum mechanics predicts the allowed states (and corresponding energies) for such particles. Now, any one of these states can be considered as 'the system', while the remaining ones altogether behave as 'the reservoir': as a matter of fact, whenever a particle moves from the selected quantum state to another one, the energy and the number of particles of the 'system' are modified to benefit the corresponding quantities of the 'reservoir'. This is precisely the situation discussed in the previous section and, therefore, we can profit from the grand canonical probabilities to calculate *the average occupation number of the quantum states of the confined gas* containing either fermions or bosons.

Let us suppose that the selected quantum state has energy E and that n particles are accommodated on it. By means of equation (3.35) the average occupation number $\langle n \rangle$ of this state is calculated as[3]

---

[2] As usual, it is understood that E_i are the single-particle quantum energy levels, while n_i is the number of particles accommodated on the level E_i.
[3] The 'system' energy is $E_s = nE$ and it is assumed that the quantum gas is in equilibrium at temperature T.

$$\langle n \rangle = \sum_{n \in \mathbb{N}} n P_s = \frac{1}{\mathcal{Z}_{GC}} \sum_{n \in \mathbb{N}} n \exp\left[-\frac{n(E - \mu_c)}{k_B T} \right] \tag{3.37}$$

where in this case the grand canonical partition function is written as

$$\mathcal{Z}_{GC} = \sum_{n \in \mathbb{N}} \exp\left[-\frac{n(E - \mu_c)}{k_B T} \right] \tag{3.38}$$

and $\mathbb{N}$ indicates the set of the integer numbers (including 0). It is useful at this stage to introduce the compact notation

$$x = \exp\left[-\frac{E - \mu_c}{k_B T} \right] \tag{3.39}$$

so that

$$\langle n \rangle = \frac{\sum_n n\, x^n}{\sum_n x^n} \tag{3.40}$$

which represents the starting point for our final discussion.

We start considering fermions, obeying the Pauli exclusion principles: no more than one particle can be accommodated on each quantum level. Therefore, in equation (3.40) we can only have $n = 1$ (fully occupied level) or $n = 0$ (empty level). Accordingly

$$\langle n \rangle = \frac{\sum_{n=0}^{1} n\, x^n}{\sum_{n=0}^{1} x^n} = \frac{x}{1 + x} = \frac{1}{\exp[(E - \mu_c)/k_B T] + 1} \tag{3.41}$$

which corresponds to equation (3.17) thus confirming the Fermi–Dirac statistics.

Let us now move to the more complicated case of bosons, whose occupation is unrestricted. In this case the denominator of equation (3.40) is the geometric series

$$\sum_{n \in \mathbb{N}} x^n = 1 + x + x^2 + x^3 + x^4 + \cdots = \frac{1}{1 - x} \tag{3.42}$$

while the numerator can be handled as follows

$$\sum_{n \in \mathbb{N}} n\, x^n = x + 2x^2 + 3x^3 + 4x^4 + \ldots = x(1 + 2x + 3x^2 + 4x^3 + \cdots) \tag{3.43}$$

which allows us to identify the term in parenthesis in the right-hand side as the first derivative of the geometric series

$$1 + 2x + 3x^2 + 4x^3 + \ldots = \frac{d}{dx}\left(\frac{1}{1 - x} \right) = (1 - x)^{-2} \tag{3.44}$$

so that

$$\langle n \rangle = \frac{x(1-x)^{-2}}{(1-x)^{-1}} = \frac{1}{\exp[(E-\mu_c)/k_BT]-1} \tag{3.45}$$

which corresponds to equation (3.30) thus confirming the Bose–Einstein statistics for a fixed number of particles.

References

[1] Sakurai J J and Napolitano J 2011 *Modern Quantum Mechanics* 2nd edn (Reading, MA: Addison-Wesley)

[2] Miller D A B 2008 *Quantum Mechanics for Scientists and Engineers* (New York: Cambridge University Press)

[3] Griffiths D J and Schroeter D F 2018 *Introduction to Quantum Mechanics* 3rd edn (Cambridge: Cambridge University Press)

[4] Bransden B H and Joachain C J 2000 *Quantum Mechanics* (Englewood Cliffs, NJ: Prentice-Hall)

[5] Colombo L 2019 *Atomic and Molecular Physics: A Primer* (Bristol: IOP Publishing)

[6] Eisberg R and Resnick R 1985 *Quantum Physics of Atoms, Molecules, Solids, Nuclei, and Particles* 2nd edn (Hoboken, NJ: Wiley)

[7] Greiner W 1990 *Relativistic Quantum Mechanics* (Berlin: Springer)

[8] Dirac P A M 1958 *The Principles of Quantum Mechanics* (Oxford: Oxford University Press)

[9] Colombo L 2021 *Solid State Physics: A Primer* (Bristol: IOP Publishing)

[10] Demtröder W 2010 *Atoms, Molecules and Photons* (Berlin: Springer)

[11] Reif F 1987 *Fundamentals of Statistical and Thermal Physics* (New York: McGraw-Hill)

[12] Kennett M P 2021 *Essential Statistical Mechanics* (Cambridge: Cambridge University Press)

IOP Publishing

Statistical Physics of Condensed Matter Systems
A primer
Luciano Colombo

Chapter 4

Thermal properties of quantum gases

Syllabus—We investigate the thermal properties of two quantum gases of paradigmatic relevance in condensed matter physics, namely the free electron gas and the phonon gas. The former describes the system of conduction electrons in a metallic solid, while the latter represents the quantum picture for the lattice dynamics of a crystal. They are governed by the Fermi–Dirac and by the Bose–Einstein statistics, respectively.

4.1 The electron gas

4.1.1 Electrons in metals

In metals it is possible to draw a very drastic approximation, in addition to the ones usually introduced to treat the physics of solid-state systems (see appendix F), namely: *we completely neglect any ion effect on electrons*. This is indeed a much stronger approximation than the adiabatic one (decoupling the ion and electron systems), since it basically states that electrons do not undergo the action of any external potential: they are considered as *free particles*. In addition, the problem is further simplified by also neglecting the internal interactions among electrons: they are further treated as *independent particles*. This assumption is only corrected by assuming that the electrons occasionally undergo collisions with other electrons, or lattice vibrations, or lattice defects. Collisions are treated as instantaneous events, occurring on average with frequency $1/\tau_e$ (where τ_e, known as the relaxation time, describes on average the time lapse between two consecutive scattering events).

The resulting picture is referred to as the *free electron gas model*: basically, the conduction electrons in a metal are treated as a gas of free and independent fermions, confined within the volume of the solid and undergoing scattering collisions. We provide a phenomenological justification for the model, starting from some robust experimental facts: (i) metals have a room-temperature resistivity in the range 10^{-8} and 10^{-6} Ω m; (ii) the presence of lattice defects detrimentally

doi:10.1088/978-0-7503-2269-0ch4

affects their charge transport properties; (iii) their resistivity decreases by decreasing temperature. The number n_e of valence electrons per cm^3 is given by the product

$$n_e = \mathcal{N}_A \, \frac{d_m}{A} \, Z_v \tag{4.1}$$

where d_m is the mass density of the metal, while the symbols $\mathcal{N}_A$, Z_v, and A are the Avogadro number, the number of valence electrons per atom (chemical valence), and the atomic mass number, respectively. Next, by attributing a spherical space volume to each electron, we can easily compare its radius $r_e = (3/4\pi n_e)^{1/3}$ with the typical interatomic distances in crystals (which are of the order of few Å) and come to the conclusion that in metals there is plenty of room available for conduction electrons. If we further take into consideration that they are only weakly bound to their ion core (since they are valence electrons), we can argue that they homogeneously delocalise throughout the interstitial regions of the solid. The typical values of these parameters characterising the conducting gas of some metals are reported in table 4.1, that hereafter we will model *as a homogeneous gas of delocalised, free, independent, and charged particles with spin $\hbar/2$*. The model is able to correctly explain the observed charge transport properties (both in direct and alternate current configuration) as well as the optical properties (namely the reflectivity and the plasma excitations) of metals [1–4]. We argue that it is similarly reliable to describe their thermal properties.

The single-particle wavefunction $\psi(\mathbf{r})$ describing the quantum states of the free electron gas is obtained by solving the eigenvalue equation

$$-\frac{\hbar^2}{2m_e}\nabla^2\psi(\mathbf{r}) = E\psi(\mathbf{r}) \tag{4.2}$$

where m_e and E are the electron mass and energy, respectively. By imposing periodic boundary conditions [1] we obtain the single-electron normalised wavefunction

$$\psi_\mathbf{k}(\mathbf{r}) = \frac{1}{\sqrt{V}}\, e^{i\mathbf{k}\cdot\mathbf{r}} \tag{4.3}$$

Table 4.1. Number Z_v of valence electrons, electron density n_e (units 10^{22}/cm^3), radius r_e (units Å) of the sphere defining the volume per electron, room-temperature resistivity ρ_e (units 10^{-8} Ω m), Fermi energy E_F (units eV), and Fermi temperature T_F (units 10^4 K) of some metallic elements.

	Li	Na	Cu	Ag	Au	Mg	Ca	Fe	Zn	Al	Sn	Pb
Z_v	1	1	1	1	1	2	2	2	2	3	4	4
n_e	4.7	2.6	8.5	5.9	5.9	8.6	4.6	17.0	13.2	18.1	14.8	13.2
r_e	1.7	2.1	1.4	1.6	1.6	1.4	1.7	1.1	1.2	1.1	1.2	1.2
ρ_e	9.5	4.9	1.7	1.6	2.2	4.4	3.4	10.0	6.0	2.7	11.5	21
E_F	4.7	3.2	7.0	5.4	5.5	7.1	4.7	11.1	9.5	11.7	10.2	9.5
T_F	5.5	3.8	8.2	6.4	6.4	8.2	5.4	13.0	11.0	13.6	11.8	11.0

where $\mathbf{k}$ is its wavevector and V is the system volume. By assuming a cubic sample with edge L, we can set $V = L^3$ so that the allowed wavevectors are spaced by $2\pi/L$ along any Cartesian direction and, therefore, their number density in the reciprocal space is $L^3/(2\pi)^3 = V/(2\pi)^3$. The corresponding electron energy is

$$E = \frac{\hbar^2 k^2}{2m_e} \tag{4.4}$$

while its momentum and velocity are $\mathbf{p} = \hbar\mathbf{k}$ and $\mathbf{v} = \hbar\mathbf{k}/m_e$, respectively.

Through equation (4.4) we understand that the electron energies only depend on the magnitude of $\mathbf{k}$ (and not on its direction in space); also, we assume that V is large, so that the energy spectrum of the electrons is practically continuous, as well as the variable k. This implies that the number of allowed wavevectors corresponding to quantum levels with energy in between E and $E + dE$ is given by the product between their number density and the infinitesimal volume $4\pi k^2 dk$, corresponding to the spherical shell in reciprocal space contained in between the two surfaces $E = $ constant and $E + dE = $ constant. According to the Pauli principle, we can place up to two paired[1] electrons on each allowed quantum level. Therefore, we define the *zero-temperature electronic density of states* (eDOS) $G(E)$ through the calculation of the quantity $G(E)dE$

$$G(E)dE = 2\,\frac{V}{(2\pi)^3}\,4\pi k^2 dk = \frac{V}{2\pi^2\hbar^3}(2m_e)^{3/2}\,E^{1/2}\,dE \tag{4.5}$$

providing the *the number of electron states with energy in the interval* $[E, E + dE]$. Since at zero temperature any state is filled up to the energy E_F (see section 3.2.2) we can impose the normalisation condition

$$\int_0^{E_F} G(E)dE = N \tag{4.6}$$

where N is the number of electrons. By inserting the eDOS provided in equation (4.5) we get

$$E_F = \frac{\hbar^2}{2m_e}(3\pi^2)^{2/3}n_e^{2/3} \tag{4.7}$$

where $n_e = N/V$ is the *electron number density*. Typical values of E_F are reported in table 4.1.

It is interesting to recognise that E_F depends on the electron number density and, therefore, is specific of any metal. In particular, the typical values of n_e are of the order of 10^{22} electrons/cm^3: a value largely exceeding the density of whatever atomic/molecular gas. This clearly indicates that the electron gas is definitely not an ordinary classical system. This statement is further supported by introducing the

---

[1] 'Paired electrons' means: electrons with opposite spin.

Fermi temperature $T_F = E_F/k_B$ which defines the intrinsic temperature scale of the conduction gas in a metal. In other words, T_F sets the temperature only above which the free electron gas could be treated classically. As reported in table 4.1 we have $T_F \sim \mathcal{O}(10^4 \text{ K})$ and, therefore, this limit is never reached.

For a continuous distribution of energy levels, the quantity $G(E)dE$ replaces the intrinsic probability p_i introduced in section 3.2.1 for a discrete energy spectrum. Accordingly, if we now move to consider the free electron gas in equilibrium at temperature $T > 0$ K, the number dn of particles with energy in the interval $[E, E + dE]$ is obtained by a straightforward generalisation of equation (3.16)

$$dn = \frac{G(E)dE}{1 + \exp[(E - \mu_c)/k_B T]} = \frac{V}{2\pi^2 \hbar^3}(2m_e)^{3/2} \frac{1}{1 + \exp[(E - \mu_c)/k_B T]} E^{1/2}\, dE \quad (4.8)$$

which in turn allows us to define the *finite-temperature electronic density of states* $G(E, T)$ as

$$G(E, T) = \frac{V}{2\pi^2 \hbar^3}(2m_e)^{3/2} \frac{1}{1 + \exp[(E - \mu_c)/k_B T]} E^{1/2} \quad (4.9)$$

whose plot is reported in figure 4.1 (thick red line), together with its zero-temperature counterpart (thin blue line). We remark that in plotting this figure we have set $\mu_c = E_F$ even if we are considering the gas at $T > 0$ K. This is tantamount *to neglecting the temperature dependence of the chemical potential*, indeed a very good approximation for the electron gas as explained in the next section.

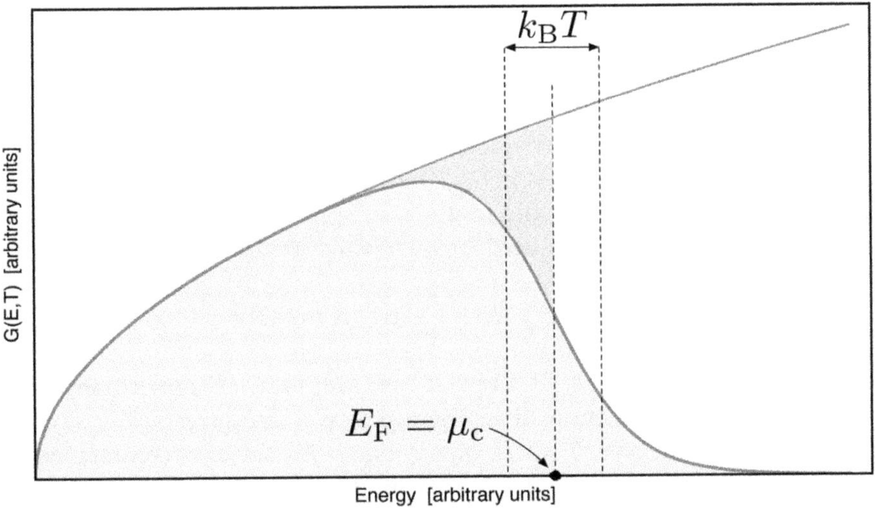

Figure 4.1. The electron density of states at zero temperature (thin blue line) and at finite temperature T (thick red line). The energy interval over which the Fermi–Dirac distribution is non constant is as large as $\sim k_B T$ and it is centred at the Fermi energy E_F which in this plot has been identified with the chemical potential (see text for explanation).

4.1.2 The chemical potential of the free electron gas

In order to validate the guess $\mu_c(T) \simeq E_F$ adopted to draw figure 4.1, we preliminarily observe that the number N of electrons is obviously unaffected by the temperature (equivalently: the total number of electrons is conserved); accordingly, we set

$$N = \int_0^{+\infty} G(E, T)\, dE = \int_0^{+\infty} G(E)f(E, T)\, dE \tag{4.10}$$

where $f(E, T)$ is the Fermi–Dirac occupation number introduced in equation (3.17). This result allows us to interpret the grey shaded area of figure 4.1 as the conserved number of electrons. More specifically, the area under the $G(E, T)$ function corresponding to the energy interval $\mu_c \leqslant E \leqslant +\infty$ represents the number of electrons promoted to energies above the Fermi energy, upon increasing temperature from zero to T. Obviously, this promotion has emptied a number of states just below E_F: the corresponding number of promoted electrons is given by the area (shaded in cyan colour in figure 4.1) in between the function $G(E, T)$ and its zero-temperature counterpart and calculated for energies such that $0 \leqslant E \leqslant E_F$.

In order to determine the actual T-dependence of the chemical potential of the free electron gas we observe that it implicitly enters in equation (4.10) through the Fermi–Dirac occupation number. An integral of the kind shown in that equation belongs to the family of *Fermi–Dirac integrals*

$$I_{FD}(E) = \int_0^{+\infty} \rho(E)f(E, T)\, dE \tag{4.11}$$

where $\rho(E)$ is any function of the sole energy E. Their calculation proceeds through the so-called *Sommerfeld expansion*, indeed a rather technical formal development that, good for us, has been already calculated once for all [5, 6]

$$I_{FD}(E) = \int_0^{\mu_c} \rho(E)\, dE + \frac{\pi^2}{6} \rho^{(1)}(\mu_c)\, (k_B T)^2 + \frac{7\pi^4}{360} \rho^{(3)}(\mu_c)\, (k_B T)^4 + \cdots \tag{4.12}$$

where $\rho^{(\xi)}(\mu_c)$ is the ξth order derivative of the $\rho(E)$ function calculated at $E = \mu_c$. This result contains a power series in $(k_B T)$ which, for our present purposes, will be truncated at the first term quadratic in T.

The Sommerfeld expansion is used to solve equation (4.10)

$$N = \int_0^{\mu_c} G(E)\, dE + \frac{\pi^2}{6} G^{(1)}(\mu_c)\, (k_B T)^2 \tag{4.13}$$

where $G^{(1)}(\mu_c) = (dG(E)/dE)_{E=\mu_c}$. Even if we are now acknowledging that the chemical potential is a function of temperature, we can nevertheless say that, as shown in figure 3.1, whenever $T \ll T_F$ we have $\mu_c \sim E_F$; since, as commented before, for any temperature of physical interest this condition is always satisfied, we can set

$$\int_0^{\mu_c} G(E)\, dE \sim \underbrace{\int_0^{E_F} G(E)\, dE}_{= \quad N} - (E_F - \mu_c)\, G(E_F) =$$
$$= \quad N \quad - (E_F - \mu_c)\, G(E_F) \tag{4.14}$$

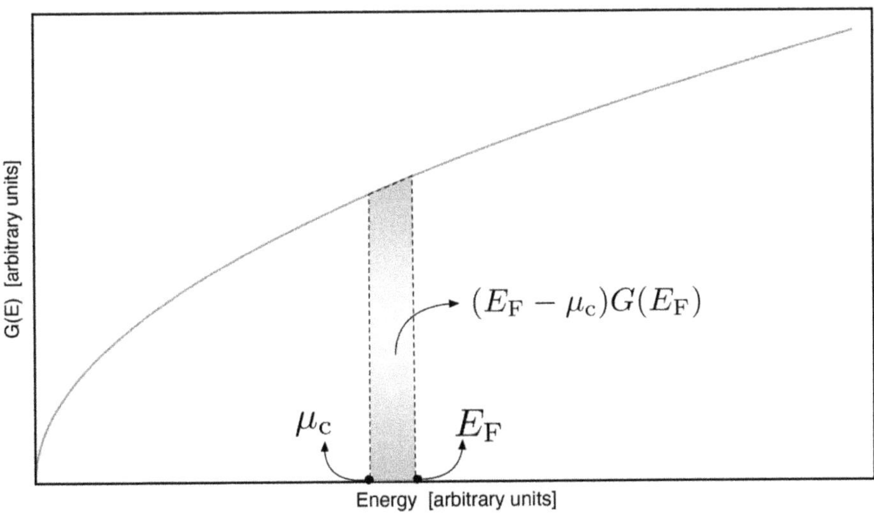

Figure 4.2. Graphical rendering of the calculation of the integral given in equation (4.14). The chemical potential and Fermi energy are indicated by μ_c and E_F, respectively. The function $G(E)$ is the eDOS for the electron gas.

where the graphical rendering of this calculations is reported in figure 4.2. By inserting this result into equation (4.13) and retaining just the first leading term of the Sommerfeld expansion we easily get

$$\mu_\mathrm{c}(T) = E_\mathrm{F}\left[1 - \frac{\pi^2}{12}\left(\frac{T}{T_\mathrm{F}}\right)^2\right] \simeq E_\mathrm{F} \tag{4.15}$$

since $T \ll T_\mathrm{F}$ as reported in table 4.1. This concludes our investigation: at any temperature of physical interest, the T-dependence of the chemical potential can be neglected.

4.1.3 Thermal properties of the free electron gas

The Sommerfeld expansion given in equation (4.12) is also useful to calculate the *internal energy of the free electron gas*

$$\mathcal{U} = \int_0^{+\infty} E\, G(E)\, f(E, T)\, dE = \mathcal{U}_0 + \frac{\pi^2}{6}\, G(E_\mathrm{F})\, (k_\mathrm{B}T)^2 \tag{4.16}$$

where

$$\mathcal{U}_0 = \int_0^{E_\mathrm{F}} E\, G(E)\, dE = \frac{V}{5\pi^2\hbar^3}(2m_\mathrm{e})^{3/2}E_\mathrm{F}^{5/2} = \frac{3}{5}NE_\mathrm{F} \tag{4.17}$$

is the zero-temperature internal energy (we made use of equation (4.7) for the Fermi energy). The non-zero quantity $\mathcal{U}_0$ provides *the minimum energy content of the fermion system* at $T = 0$ K.

This is an important result, with relevant consequences. At first, we remark that the *the average energy per electron* at $T = 0$ K is

$$\langle E \rangle = \frac{\mathcal{U}_0}{N} = \frac{3}{5}E_{\mathrm{F}} \tag{4.18}$$

in striking contrast with classical physics which would predict electrons to have zero energy. Next, by means of the first law of thermodynamics given in equation (1.36) (see also appendix C) we calculate *the pressure of the electron gas* at zero temperature

$$P_0 = -\frac{\partial \mathcal{U}_0}{\partial V} = \frac{2}{5}\frac{N}{V}E_{\mathrm{F}} \tag{4.19}$$

which, even in this case, results as non-zero as a direct consequence of the Pauli principle imposing that in a fermion system the kinetic energy is never zero. These features clearly indicate that the physics of the electron gas is inherently quantum, as further proved by computing the ratio between the mean square velocity and the square of the mean velocity of the electrons. A tedious calculation provides $\langle v^2 \rangle / (\langle v \rangle)^2 = 16/15$, while for a classical gas this ratio is $3\pi/8$: in other words, the free electron gas does not follow the Maxwell law for the velocity distribution[2].

The knowledge of the internal energy given in equation (4.17) allows for the calculation of the *constant-volume specific heat* $c_V^{\mathrm{e}}(T)$ of the free electron gas

$$c_V^{\mathrm{e}}(T) = \frac{1}{V}\frac{\partial \mathcal{U}}{\partial T}\bigg|_V = \frac{1}{V}\frac{\pi^2}{3}\,G(E_F)\,k_{\mathrm{B}}^2\,T = \frac{3}{2}n_e k_{\mathrm{B}}\left(\frac{1}{3}\pi^2\frac{T}{T_{\mathrm{F}}}\right) \tag{4.20}$$

predicting a linear T-dependence, fully consistently with experimental data [1–3]. By comparing this expression with the result $c_V^{\mathrm{e}}(T) = 3n_e k_{\mathrm{B}}/2$ provided by a classical treatment of the electron gas[3], we understand that quantum mechanics acts as a correction to the classical result, predicting a lower specific heat than classical physics, once again in full agreement with experimental observations. Anyway, we stress that the electron contribution to the total specific heat of a metallic solid is proportional to T/T_{F} which, at any temperature of physical interest, makes it really marginal as compared to the lattice (phonon) contribution described in the next chapter.

We finally address the *thermal transport* in the electron gas. The temperature-dependent parameter describing the ability of a system to transport heat is known as *thermal conductivity* $\kappa(T)$. A general result of kinetic theory [7] is that

$$\kappa(T) = \frac{1}{3}\langle \tau \rangle \langle v^2 \rangle c_V(T) \tag{4.21}$$

where $\langle \tau \rangle$ and $\langle v^2 \rangle$ are the relaxation time and the mean square velocity of the heat carriers, respectively, while $c_V(T)$ is the constant-volume specific heat of the system.

---

[2] This result is of course expected since the Maxwell law follows from the Boltzmann statistics.
[3] Basically, we apply the equipartition of energy to electrons, as if they were classical particles.

This result will be derived explicitly in the case of the phonon gas in section 4.2.3. Here we assume that equation (4.21) is as well valid for the electron gas: an heuristic argument will be offered soon to support our guess. More specifically, we set $\langle \tau \rangle = \tau_e$, namely we identify the relaxation time with the average time interval between two consecutive electron scattering events[4]. As for $\langle v^2 \rangle$ we simply take the square of the Fermi velocity $v_F = \hbar k_F/m_e$, where the Fermi wavevector is defined as $k_F = (2m_e E_F/\hbar^2)^{1/2}$. Finally, the specific heat is provided by equation (4.20). In conclusion, we get the *thermal conductivity of the free electron gas*

$$\kappa_e(T) = \frac{1}{3}\tau_e \, v_F^2 \, \frac{3}{2}n_e k_B \left(\frac{1}{3}\pi^2\frac{T}{T_F}\right) \tag{4.22}$$

which can be used together with the standard Drude expression for the electrical conductivity $\sigma_e = n_e e^2 \tau_e/m_e$ to predict the value of the Lorenz number[5]

$$L = \frac{\kappa_e(T)}{\sigma_e T} = 2.44 \times 10^{-8} \text{ W } \Omega \text{ K}^{-2} \tag{4.23}$$

indeed in excellent agreement with the value 2.41×10^{-8} W Ω K^{-2} provided by the experimental Wiedemann–Franz law, reporting that in most metals the ratio between the thermal and electrical conductivities is proportional to the temperature. This agreement between the theory here developed and the experimental evidence supports our approach to the heat transport in the electron gas through equation (4.21).

4.2 The phonon gas

4.2.1 Lattice vibrations

The crystal structure consists in an ordered array of ions regularly placed at positions $\mathbf{R}(l)$, where l is an index spanning a discrete periodic lattice[6] (see appendix F). Assuming that they remain clamped at their crystal positions is inconsistent with a body of empirical evidence, including (but not limited to) the fact that solids expand by increasing the temperature, that they can absorb infrared radiation, that they transmit heat, and that they sustain the propagation of compressive mechanical waves (sound). Therefore, we must admit that *ions vibrate around their mean positions*, as indeed proved by x-ray scattering measurements [1, 8]: the underlying crystal structure does clearly appear in the experimental diffractograms, in spite of

---

[4] These scattering events are the microscopic source of the thermal resistivity. Calculating τ_e is a specific task of solid-state theory [1–4].

[5] The relaxation times ruling over charge transport and heat transport have been here considered the same. While this is a good approximation, greater detail is required to develop a more fundamental and rigorous transport theory, as explained elsewhere [1, 3, 8].

[6] In this approach to the fundamentals of lattice dynamics, we will make use of a simplified notation, where there is no distinction between the crystal lattice and its basis. Accordingly, there is a unique label identifying each lattice site.

the fact that ions undergo movement during diffraction. This provides robust evidence that the amplitude of the ionic motion is not so large as to alter the underlying lattice structure.

We name $\mathbf{u}(l)$ the displacement of the ion in position $\mathbf{R}(l)$ and we assume it is small enough to expand the vibrational potential energy U_{vib} of the crystal in a power series truncated just to the first non-vanishing term (this is usually referred to as the *harmonic approximation*)

$$U_{\text{vib}} = \frac{1}{2}\sum_{\substack{l,m \\ \alpha\beta}} \frac{\partial^2 U}{\partial u_\alpha(l)\partial u_\beta(m)}\bigg|_0 u_\alpha(l)u_\beta(m) = \frac{1}{2}\sum_{\substack{l,m \\ \alpha\beta}} U_{\alpha\beta}(lm)\, u_\alpha(l)u_\beta(m) \qquad (4.24)$$

where the Greek indices label the Cartesian components, while the Latin indices run over the lattice sites; equation (4.24) provides the energy of a set of classical harmonic oscillators with force constants $U_{\alpha\beta}(lm)$ given by the second derivatives of the energy with respect to the ionic displacements, calculated in the rest condition. In this approximation, the ionic equations of motion are

$$M_l \ddot{u}_\alpha(l) = -\sum_{\beta m} U_{\alpha\beta}(lm)u_\beta(m) \qquad (4.25)$$

where M_l is the mass of the ion in position $\mathbf{R}(l)$. By solving these set of equations according to the standard methods of the theory of lattice dynamics [1–4], we obtain the frequencies $\omega_s(\mathbf{q})$ of the crystalline *vibrational normal modes*; here, $\mathbf{q}$ is the mode wavevector, while the label s defines its optical/acoustic character as well as its longitudinal/transverse polarisation. In conclusion, within the harmonic approximation the classical description of the lattice dynamics of a crystal is provided by calculating the dispersion relations $\omega = \omega_s(\mathbf{q})$ of its normal modes of vibration.

The switch to a quantum picture is obtained straightforwardly by *describing each classical* (*s*$\mathbf{q}$) *vibrational mode as a quantum one-dimensional harmonic oscillator* [9–11] whose energy is restricted to the values

$$E_{s\mathbf{q}} = \left(\lambda_{s\mathbf{q}} + \frac{1}{2}\right)\hbar\omega_s(\mathbf{q}) \qquad (4.26)$$

where $\lambda_{s\mathbf{q}} = 0, 1, 2, \ldots$ is the *vibrational quantum number*. Therefore, $E_{s\mathbf{q}}$ is the energy of a single $s\mathbf{q}$ quantum oscillator in its $\lambda_{s\mathbf{q}}$ th excited state. However, since the vibrational energy levels are equally spaced, the very same quantity can equivalently be looked at as the *energy of $\lambda_{s\mathbf{q}}$ identical oscillators with the same frequency $\omega_s(\mathbf{q})$*. We will adopt this second approach since it is especially effective in describing the thermal characteristics of a crystal lattice through the properties of its *gas of phonons*, namely a gas of pseudo-particles introducing the quantum corpuscular description of lattice dynamics in terms of the quanta of the ionic displacement field. *Phonons obey the Bose–Einstein statistics*, since they are identical and indistinguishable quantum objects with zero spin.

It must be understood that phonons, similarly to photons, are named *pseudo-particles* since they do not have a mass, nevertheless they carry a momentum $\hbar\mathbf{q}$ and have group velocity $d\omega_s(\mathbf{q})/d\mathbf{q}$. Since their momentum is exact, the uncertainty principle imposes that the phonon position is totally undetermined and, therefore, they must be considered as *delocalised pseudo-particles*. This is consistent with the fact that their corresponding non-interacting classical vibrational modes extend throughout the system.

Finally, we must take into consideration that *the total number of phonons is not conserved*. For instance, they can be created or annihilated by increasing or decreasing the crystal temperature. It is easy to understand this by starting from the classical picture: increasing the temperature makes the ions oscillate with larger amplitudes. Accordingly, each normal mode $s\mathbf{q}$ has larger intensity or, equivalently, its energy is increased. Within the quantum harmonic theory this corresponds to an increase of the vibrational quantum number $\lambda_{s\mathbf{q}}$ or, ultimately, to an increase of the number of $s\mathbf{q}$ phonons. The inverse situation is found by cooling the crystal, a very effective phonon-annihilation process indeed. Crystal defects also affect the number of phonons, since they act as scattering centres where creation or annihilation events may occur. Also, by developing the theory of lattice dynamics beyond the harmonic approximation, it is possible to prove that anharmonic terms in the vibrational potential energy are associated in the quantum picture with phonon–phonon scattering events [12].

Since the phonon number is not conserved, we must readdress the formalism developed in section 3.3. More specifically, the condition $\sum_i dn_i = 0$ can be dropped and, accordingly, in searching for the equilibrium distribution there is no need to introduce the Lagrange multiplier α to maximise the quantity $-d \ln P$. This leads to a version of the *Bose–Einstein occupation number for phonons*

$$n(s\mathbf{q}, T) = \frac{1}{\exp[\hbar\omega_s(\mathbf{q})/k_\mathrm{B}T] - 1} \tag{4.27}$$

where the index i appearing in equation (3.30) has been replaced by $(s\mathbf{q})$ since both the label s and the waveverctor $\mathbf{q}$ are needed to specify the energy of each phonon. In appendix G we provide an alternative calculation leading to the same result reported in equation (4.27) and providing a phenomenological derivation (in this specific case) of the Bose–Einstein distribution law.

In conclusion, $n(s\mathbf{q}, T)$ represents *the average number of phonons with energy $\hbar\omega_s(\mathbf{q})$ found in the crystal in equilibrium at temperature T*: it may be affected by increasing/decreasing the temperature, as well as by scattering events (among phonons or by defects or by electrons).

4.2.2 The specific heat of the phonon gas

In the equilibrium condition at temperature T, the internal energy $\mathcal{U}$ of the phonon gas is

$$\mathcal{U} = \sum_{s\mathbf{q}} (n_{s\mathbf{q}} + 1/2)\hbar\omega_s(\mathbf{q}) \tag{4.28}$$

where the energy content $\mathcal{U}_0$ of the static lattice has been set to zero for convenience[7]. The constant-volume heat capacity of the phonon gas is accordingly calculated as

$$C_V = \frac{\partial}{\partial T}\sum_{s\mathbf{q}}[n(s\mathbf{q}, T) + 1/2]\hbar\omega_s(\mathbf{q}) = \sum_{s\mathbf{q}}\hbar\omega_s(\mathbf{q})\frac{\partial n(s\mathbf{q}, T)}{\partial T} = \sum_{s\mathbf{q}}c_V^{(s\mathbf{q})}(T) \quad (4.29)$$

where we have introduced the single $s\mathbf{q}$ mode heat capacity $c_V^{(s\mathbf{q})}(T)$. This expression clearly indicates a temperature dependence which is not obtained by a classical approach, where each lattice ion is treated as a three-dimensional classical oscillator and equipartition of energy is imposed. Following these assumptions, we calculate a lattice constant-volume heat capacity $C_V = 3R$, a result known as Dulong–Petit law. In contrast, the full quantum picture we have elaborated predicts that $C_V \sim T^3$ for vanishingly small temperatures, just in perfect agreement with experimental data. In figure 4.3 we report a direct comparison between the Dulong–Petit law and the actual specific heat predicted by equation (4.29). In calculating the quantum specific heat the electron contribution has been neglected (see section 4.1.3) and, therefore, figure 4.3 corresponds to the case of an insulating material where C_V is largely provided by lattice vibrations only or, equivalently, by phonons.

It is worth stressing that the excellent agreement between equation (4.29) and experimental data is not only due to the quantum treatment of lattice vibrations, but also to the complete description of the vibrational modes though their dispersion relations $\omega = \omega_s(\mathbf{q})$. Indeed, we might have been tempted to follow an intermediate approximation between the Dulong–Petit model and the full quantum treatment,

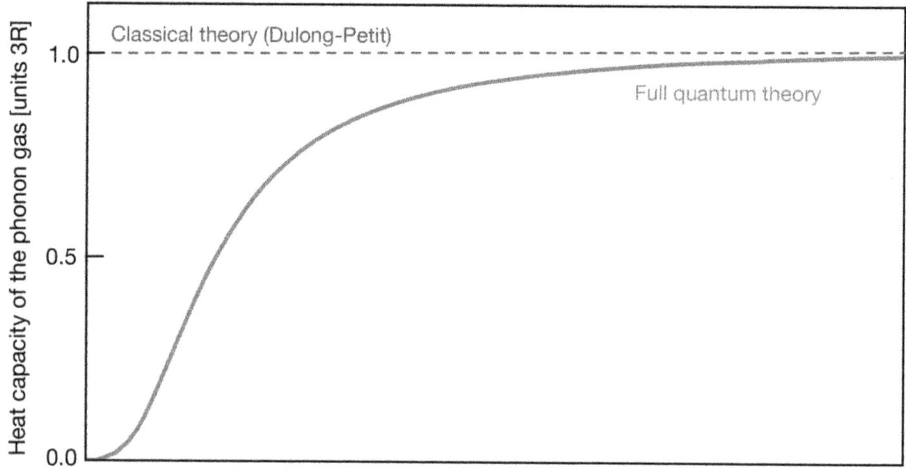

Figure 4.3. The lattice heat capacity predicted by the quantum phonon gas model (full red line) and by the classical Dulong–Petit law (dashed blue line). As $T \to 0$ K the specific heat drops to zero as T^3.

---

[7] This is motivated by the fact that $\mathcal{U}_0$ does not depend on temperature [1].

assuming, for example, that the oscillators were really quantum, but all with equal frequency ω_E: this is called the Einstein model. The average energy of a single one-dimensional Einstein oscillator at temperature T is

$$\langle u_E \rangle = \left[\frac{1}{\exp(\hbar\omega_E/k_B T) - 1} + \frac{1}{2} \right] \hbar\omega_E \tag{4.30}$$

and the corresponding *Einstein heat capacity* is calculated as

$$C_V^{\text{Einstein}} = 3R \left(\frac{\hbar\omega_E}{k_B T} \right)^2 \frac{\exp(\hbar\omega_E/k_B T)}{[\exp(\hbar\omega_E/k_B T) - 1]^2} \tag{4.31}$$

an expression which describes a T-dependence $C_V^{\text{Einstein}} \to \exp(-1/T)$ at vanishingly small temperatures, in disagreement with experimental data. A possible correction to the Einstein model is provided by the Debye model, where phonon dispersion relations are indeed taken into account, but they are approximated to just three effective acoustic branches. While we direct the reader to any good textbook of solid-state physics [1–3], here we simply recall that the Debye model predicts a correct trend for C_V at small temperature, but it is less accurate than the full quantum theory here developed; its convenience relies on the fact that it does not require the full calculation of phonon dispersion relations.

4.2.3 The thermal conductivity of the phonon gas

Within the harmonic approximation discussed in the previous sections, phonons are described as a gas of free pseudo-particles. Actually, they undergo mutual interactions because of the terms neglected in expanding the potential energy of the crystal as a power series of the ionic displacements (see equation 4.24). In other words, whenever we go beyond the harmonic approximation we must admit that phonon–phonon interactions do occur: third-order terms introduce three-phonon interactions, fourth-order terms introduce four-phonon interactions, and so on [1–3]. However, the harmonic vibrational states of the crystal remain basically unaffected by phonon–phonon interactions since *anharmonicity acts as a perturbation, only promoting transitions between different states of quantum harmonic oscillator.* We can therefore still speak about phonon frequencies and vibrational modes with different character s and wavevector $\mathbf{q}$. The rigorous treatment of such a perturbation is not at all trivial [12, 13], falling beyond the level of the present formalism, but the underlying physical picture is in fact simple: we can pictorially say that the nth order term (with $n \geqslant 3$) in the Taylor expansion of the vibrational potential energy activates interactions among n phonons, which we will refer to as *n-phonon scattering events*. Since the phonon number is not conserved, during a scattering event phonons of some harmonic modes are annihilated (their population is decreased), while other phonons of different modes are created (their population is increased). Any such scattering event is characterised by its own rate of occurrence per unit time, by means of which[8] we can calculate the resulting average lifetime $\tau_{s\mathbf{q}}$ for each $s\mathbf{q}$ vibrational mode or, equivalently, for each phonon.

In order to investigate the heat transport in a solid-state system, let us consider a homogeneous insulating[9] crystal across which a small temperature gradient is imposed. The steady state thermal conduction regime is described by the phenomenological Fourier law [14, 15]

$$J_{h,x} = -\kappa_1 \frac{dT}{dx} \tag{4.32}$$

where $J_{h,x}$ is the heat flux along x, namely the amount of thermal energy crossing in the unit time a unit area (it is measured in units $\mathrm{J\,m^{-2}\,s^{-1}}$) normal to the x direction of the applied thermal gradient. We aim at developing a microscopic theory of the *lattice thermal conductivity* κ_1 which is the material-specific quantity making the difference between thermal insulators (low κ_1 values) and good thermal conductors (large κ_1 values). By assuming that phonons are the only microscopic heat carriers, the heat flux $J_{h,x}$ can be calculated by summing over all vibrational $s\mathbf{q}$ modes the product heat flux = [energy carried] $\times$ [number of phonons] $\times$ [speed of propagation] $\times$ [$1/V$] where V is the volume crossed by the thermal energy flux, or equivalently

$$J_{h,x} = \frac{1}{V} \sum_{s,q} \hbar\omega_s(\mathbf{q}) \, \bar{n}(s\mathbf{q}, T) \, v_{g,x}(s\mathbf{q}) \tag{4.33}$$

where it must be noted that *the phonon population $\bar{n}(s\mathbf{q}, T)$ is not given by the Bose–Einstein distribution* since the condition here considered is not an equilibrium one.

Upon removal of the temperature gradient, the system recovers its equilibrium situation after a suitable time. This implies that during such a transient regime each single-mode phonon population $\bar{n}(s\mathbf{q}, T)$ must relax back to its equilibrium value $n(s\mathbf{q}, T)$. We guess that

$$\frac{\partial \bar{n}(s\mathbf{q}, T)}{\partial t} = -\frac{\bar{n}(s\mathbf{q}, T) - n(s\mathbf{q}, T)}{\tau_{s\mathbf{q}}} \tag{4.34}$$

where the single-mode relaxation time $\tau_{s\mathbf{q}}$ is identified with the lifetime of the corresponding vibrational mode, as discussed in the previous section. Equation (4.34) is known as the *single-mode relaxation time approximation* (SM-RTA). Basically it states that each phonon mode relaxes independently of the others.

Let us now consider a unit volume element V within the solid and name $\mathcal{U}$ its internal energy given by equation (4.28), where it is used the non-equilibrium

---

[8] Let us suppose that the $s\mathbf{q}$ phonon is affected by many possible anharmonic scattering mechanisms with rates $P_{s\mathbf{q}}^1$, $P_{s\mathbf{q}}^2$, $P_{s\mathbf{q}}^3$, ..., respectively. Then, its lifetime is calculated as $1/\tau_{s\mathbf{q}} = \sum_i P_{s\mathbf{q}}^i$. This result known as Matthiessen rule and it holds as well for any other scattering mechanism, like for instance phonon scattering by lattice defects.

[9] Considering an insulating solid allows for neglecting the electron contribution to thermal transport.

population $\bar{n}(s\mathbf{q}, T)$. In steady state conditions we can write a continuity equation for the thermal current

$$\frac{1}{V}\frac{\partial \mathcal{U}}{\partial t} = \frac{\partial J_{\mathrm{h},x}}{\partial x} \tag{4.35}$$

where no term describing the rate of energy generation/removal appears since by choice the selected unit volume does not contain any energy source or sink. By inserting equations (4.28) and (4.33) into equation (4.35) we easily obtain

$$\frac{\partial \bar{n}(s\mathbf{q}, T)}{\partial t} - v_{g,x}(s\mathbf{q})\,\frac{\partial \bar{n}(s\mathbf{q}, T)}{\partial T}\,\frac{\partial T}{\partial x} = 0 \tag{4.36}$$

where we will further assume same temperature dependence in the non-equilibrium population and in the Bose–Einstein distribution

$$\frac{\partial \bar{n}(s\mathbf{q}, T)}{\partial T} = \frac{\partial n(s\mathbf{q}, T)}{\partial T} \tag{4.37}$$

so that

$$\bar{n}(s\mathbf{q}, T) = n(s\mathbf{q}, T) - \tau_{s\mathbf{q}}\, v_{g,x}(s\mathbf{q})\,\frac{\partial n(s\mathbf{q}, T)}{\partial T}\,\frac{\partial T}{\partial x} \tag{4.38}$$

a result referred to as *linearised Boltzmann transport equation* (BTE), allowing for a straightforward calculation of the non-equilibrium phonon population $\bar{n}(s\mathbf{q}, T)$.

We can now calculate the heat flux

$$
\begin{aligned}
J_{\mathrm{h},x} &= \frac{1}{V}\sum_{s\mathbf{q}} \hbar\omega_s(\mathbf{q})\, n(s\mathbf{q}, T)\, v_{g,x}(s\mathbf{q}) \\
&\quad - \frac{1}{V}\sum_{s\mathbf{q}} \hbar\omega_s(\mathbf{q})\tau_{s\mathbf{q}}\, v_{g,\,x}^2(s\mathbf{q})\,\frac{\partial n(s\mathbf{q}, T)}{\partial T}\,\frac{\partial T}{\partial x}
\end{aligned} \tag{4.39}
$$

where we have inserted the non-equilibrium population provided by BTE into equation (4.33). The first term on the right hand side does not contribute to the heat current since it only depends on equilibrium quantities. Therefore, by comparing equations (4.32) and (4.39) we eventually obtain a *microscopic equation for the lattice thermal conductivity*

$$
\begin{aligned}
\kappa_{\mathrm{l}}(T) &= \frac{1}{V}\sum_{s\mathbf{q}} \hbar\omega_s(\mathbf{q})\, \tau_{s\mathbf{q}}\, v_{g,\,x}^2(s\mathbf{q})\,\frac{\partial n(s\mathbf{q}, T)}{\partial T} \\
&= \frac{1}{3V}\sum_{s\mathbf{q}} \tau_{s\mathbf{q}}\, \langle v_{g,\,s\mathbf{q}}^2\rangle\, C_{V,s\mathbf{q}}(T) \\
&= \frac{1}{3}\,\langle\tau\rangle\,\langle v_g^2\rangle\, c_V(T)
\end{aligned} \tag{4.40}
$$

which has been presented in the three more commonly used forms. More specifically, the second form is obtained by defining a mode-specific scalar mean square group

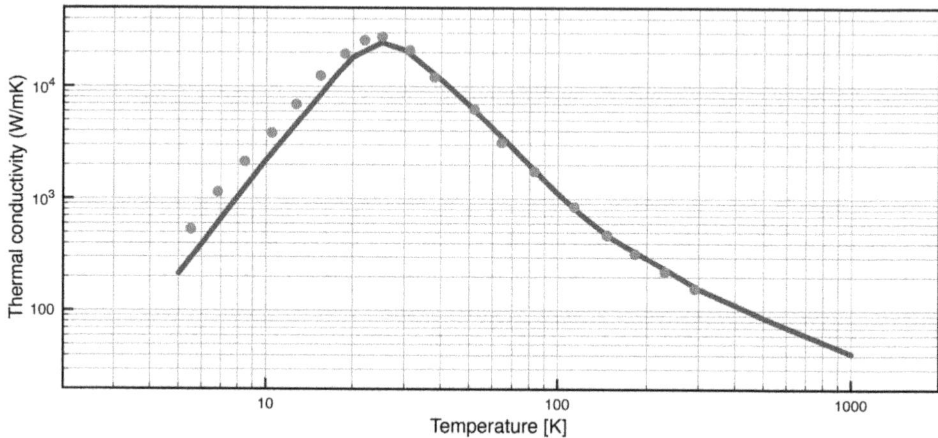

Figure 4.4. Full blue line: the lattice thermal conductivity of crystalline silicon calculated by the SM-RTA-BTE approach. The calculations have been performed by using the QUANTUM ESPRESSO integrated suite of Open-Source computer codes for materials modelling; see website https://www.quantum-espresso.org (by courtesy of Giorgia Fugallo). Red dots: experimental data taken from [16], copyright IOP Publishing, all rights reserved.

velocity $\langle v_{g,\,sq}^2 \rangle/3 = v_{g,\,x}^2(s\mathbf{q}) = v_{g,\,y}^2(s\mathbf{q}) = v_{g,\,z}^2(s\mathbf{q})$, while the third form is found by further attributing the same mean value of relaxation time $\langle \tau \rangle$ and square group velocity $\langle v_g^2 \rangle$ to all phonon modes, as well as by using the constant-volume specific heat $c_V(T) = (1/V)\sum_{sq} \mathcal{C}_{V,sq}(T) = \mathcal{C}_V(T)/V$. This latter form has been previously used in equation (4.21) to evaluate the thermal conductivity of the free electron gas. The basic reason why it holds for phonons as well as for electrons is that the very same expression is provided by the general elementary kinetic theory [7] for whatever kind of heat carriers.

In figure 4.4 we report the thermal conductivity of crystalline silicon calculated according to the SM-RTA-BTE theory. The agreement with experimental data (red dots) is impressive, proving that the adopted hierarchy of approximations is rather physically sound. On this basis we argue that the picture we elaborated in this section is generally valid in most non-metallic crystals when only phonon-mediated heat currents are observed.

Finally, it is interesting to observe that in the harmonic approximation there are no phonon–phonon interactions and, therefore, the lifetime τ_{sq} of any vibrational mode is infinite. According to equation (4.40), this would suggest an infinite lattice thermal conductivity: indeed striking failure of the purely harmonic crystal picture.

References

[1] Colombo L 2021 *Solid State Physics: A Primer* (Bristol: IOP Publishing)
[2] Kittel C 1996 *Introduction to Solid State Physics* 7th edn (Hoboken, NJ: Wiley)
[3] Ashcroft N W and Mermin N D 1976 *Solid State Physics* (London: Holt-Saunders International Editions)

[4] Grosso G and Pastori Parravicini G 2014 *Solid State Physics* 2nd edn (Oxford: Academic)

[5] Kennett M P 2021 *Essential Statistical Mechanics* (Cambridge: Cambridge University Press)

[6] Swendsen R H 2012 *An Introduction to Statistical Mechanics and Thermodynamics* (Oxford: Oxford University Press)

[7] Reif F 1987 *Fundamentals of Statistical and Thermal Physics* (New York: McGraw-Hill)

[8] Hook J R and Hall H E 2010 *Solid State Physics* (Hoboken, NJ: Wiley)

[9] Miller D A B 2008 *Quantum Mechanics for Scientists and Engineers* (New York: Cambridge University Press)

[10] Bransden B H and Joachain C J 2000 *Quantum Mechanics* (Englewood Cliffs, NJ: Prentice-Hall)

[11] Sakurai J J and Napolitano J 2011 *Modern Quantum Mechanics* 2nd edn (Reading, MA: Addison-Wesley)

[12] Böttger H 1983 *Principles of the Theory of Lattice Dynamics* (Berlin, DDR: Akademie-Verlag)

[13] Srivastava G P 1990 *The Physics of Phonons* (Bristol: Adam Higler)

[14] Incoprera F P and Dewitt D P 2011 *Fundamentals of Heat and Mass Transfer* (New York: Wiley)

[15] Lienhard H H IV and Lienhard J H V 2006 *A Heat Transfer Textbook* (Cambridge, MA: Phlogiston Press)

[16] Inyushkin A V, Taldenkov A N, Gibin A M, Gusev A V and Pohl H-J 2004 On the isotope effect in thermal conductivity of silicon *Phys. Status Solidi (c)* **1** 2995–8

IOP Publishing

Statistical Physics of Condensed Matter Systems
A primer
Luciano Colombo

Chapter 5

Other quantum systems and phenomena

Syllabus—*We conclude our journey into the realm of quantum statistical physics by investigating at first two new boson systems of special relevance in condensed matter physics, namely: the photon gas and the quantum ideal particle gas. In the first case, we derive the spectral energy density for the blackbody radiation. While this result was originally obtained at the dawn of quantum mechanics by Planck through heuristic arguments, here we derive it by directly applying the Bose–Einstein statistics. Next, we calculate the internal energy of an ideal boson gas with fixed number of particles, proving that quantum features are just second order corrections to the classical theory. Eventually, we present a simple approach to Bose–Einstein condensation, only observed in physical systems not subject to the Pauli exclusion principle.*

5.1 The photon gas

The capacity of a solid body to absorb and emit electromagnetic radiation at any finite temperature is described by its spectral absorption and spectral emission coefficients, respectively, defined as the electromagnetic power absorbed or emitted at any given frequency by their unit of surface. A blackbody is an ideal system characterised by a spectral absorption as large as 100%. In practice, it can be mimicked by a solid sample containing a cavity. Each vibrating atom of the cavity walls behaves as a radiator emitting an electromagnetic wave at the same frequency it oscillates. The field so generated, referred to as *blackbody radiation*, is trapped into the cavity and, in equilibrium conditions, the energy distribution of the atomic radiators and of the blackbody radiation must be just the same[1].

According to the semi-classical description [1, 2], *a radiation field of frequency ν consists of light quanta called photons*, which are described as massless and zero-spin pseudo-particles, each carrying an anergy $E = h\nu$ and a momentum $p = h/\lambda$ (where $\lambda = c/\nu$ is the photon wavelength and c is the speed of light). According to this

---

[1] If not, a net energy flux should occur between matter and radiation.

doi:10.1088/978-0-7503-2269-0ch5

picture, *the blackbody radiation is described as a photon gas at thermal equilibrium,* whose pseudo-particles do not interact among themselves (they are free), but they can be absorbed/emitted by the cavity walls: their number is therefore non-conserved. In short, the photon gas describing the blackbody radiation must obey the Bose–Einstein statistics, with no constraint to conserve the particle number. In addition, the photon energy spectrum can be treated as continuous[2]. Under these conditions, the number dn of photons with energy in between E and $E + dE$ is given by the continuous counterpart of equation (3.29)

$$dn = \frac{G(E)dE}{\exp(E/k_{\mathrm{B}}T) - 1} \tag{5.1}$$

where we have set $\alpha = 0$ and $G(E)dE$ is the number of photon states in the same range.

Since for photons $E = h\nu$, we may set $G(E)dE = G(\nu)d\nu$ where the right-hand side of this equation provides the number of oscillatory modes of the blackbody radiation with frequency in between ν and $\nu + d\nu$. If the electromagnetic field is trapped into a cavity with volume V, then it is calculated (see appendix H) that

$$G(\nu)d\nu = \frac{8\pi V}{c^3}\nu^2 d\nu \tag{5.2}$$

so that

$$dn = \frac{8\pi V}{c^3}\frac{\nu^2}{\exp(h\nu/k_{\mathrm{B}}T) - 1}d\nu \tag{5.3}$$

leading to the *blackbody spectral energy density*

$$u_{\mathrm{BB}}(\nu) = \frac{1}{V}h\nu\frac{dn}{d\nu} = \frac{8\pi h}{c^3}\frac{\nu^3}{\exp(h\nu/k_{\mathrm{B}}T) - 1} \tag{5.4}$$

which is known as the *Planck equation* for the blackbody radiation[3]. Equation (5.4) is in excellent agreement with experimental data, thus confirming the robustness of the statistical approach we developed. More specifically, by creating a small orifice in the cavity, it is possible to directly measure the spectral distribution of the emitted blackbody energy density: at any temperature, it is found to be exactly the

---

[2] More precisely, the photon energy spectrum is continuous provided that the dimension of the cavity trapping the blackbody radiation is larger than the typical photon wavelength, so that the energy distance between two consecutive confined modes is actually very small.

[3] We remark that the original derivation of the Planck equation was based on a rather different approach, namely by calculating the energy density distribution of the atomic radiators in the cavity walls. This derivation, is no longer considered as physically sound since it was based on two wrong assumptions: (i) the energy of the atomic radiators, described as quantum oscillators, was set as $E_{\mathrm{vib}} = nh\nu$, contrary to rigorous quantum mechanics, and (ii) the Boltzmann statistics was used for them. In this respect, the original derivation is fortuitous. Nevertheless, the original Planck argument was seminal in assuming that the interaction between electromagnetic radiation and matter occurs through emission/absorption of energy quanta as large as $h\nu$. The Planck equation represents one of the major achievements of early quantum mechanics.

non-monotonic function of the frequency predicted by the Planck equation[4]. As for the temperature dependence of $u_{BB}(\nu)$, the experimental findings are summarised by two laws: the total blackbody electromagnetic power $P_{BB}(T)$ emitted by the unit area is found proportional to the fourth power of T

$$P_{BB}(T) = \sigma_S T^4 \tag{5.5}$$

where $\sigma_S = 5.67 \times 10^{-8}$ W m^{-2} K^{-4} is known as Stefan constant and it holds

$$\frac{\nu_{max}}{T} = \text{constant} \tag{5.6}$$

where ν_{max} is the frequency at which the maximum emission intensity is observed. This result is usually referred to as the Wien displacement law. Both experimental laws are easily obtained from equation (5.4), by means of which we can in particular predict a value of the Stefan constant in pretty good agreement with experimental data.

Among many other applications of the Planck law, it is worth mentioning that the accurately measured uniform cosmic microwave background radiation (which is a reminiscence of the primordial history of the Universe, confirming the hypothesis of its birth from a 'big bang' event [3]) strictly follows the blackbody spectrum. From this, it has been estimated a nearly-uniform universe temperature as low as 2.7 K: this is the cosmological thermal bath.

5.2 The quantum ideal gas

We have investigated the ideal gas model in section 1.7 using classical statistical physics. In fact, at the most fundamental level all particles are described either as fermions or as bosons and, therefore, a proper quantum statistics should be duly developed for such a system. Since most molecules have integer (or zero) spin, it seems appropriate to develop a theory for the *quantum boson gas*[5]. This model system is important also for another reason: at variance with the phonon and photon gases, it represents a boson system *with conserved number of particles*. Accordingly, the α Lagrange multiplier entering the Bose–Einstein statistics (see discussion developed in section 3.3) must be explicitly taken into consideration and determined.

In an ideal gas the particles are distributed on a continuous energy spectrum and, therefore, equation (3.29) must be replaced by its continuous counterpart

$$dn = \frac{G(E)dE}{\exp(\alpha)\exp(E/k_BT) - 1} = \frac{4\pi V}{h^3}(2m^3)^{1/2}\frac{E^{1/2}}{\exp(\alpha)\exp(E/k_BT) - 1}dE \tag{5.7}$$

---

[4] In the Planck original derivation the actual value of the constant h was not known: in fact, it was determined by fitting equation (5.4) on the experimental curve. Remarkably, the fitted h value is just the same at any temperature.

[5] Ideal gases following the Fermi–Dirac statistics are treated elsewhere [4–7].

where V is the gas volume and $G(E)dE$ indicates the number of molecules with energy in the interval $[E, E + dE]$, for which we used the density of energy levels given in equation (1.50). The number of molecules N is calculated as

$$N = \frac{2\mathcal{Z}}{\sqrt{\pi}} \int_0^{+\infty} \frac{\zeta^{1/2}}{\exp(\alpha)\exp(\zeta) - 1} d\zeta \tag{5.8}$$

where $\zeta = E/k_B T$ and $\mathcal{Z}$ is the partition function of the ideal gas, given in equation (1.52). This equation can be expanded as[6]

$$N = \mathcal{Z}\exp(-\alpha)[1 + 2^{-3/2}\exp(-\alpha) + \ldots] \tag{5.9}$$

from which we obtain the Lagrange multiplier α

$$\exp(-\alpha) = \frac{N}{\mathcal{Z}}\left(1 - 2^{-3/2}\frac{N}{\mathcal{Z}} + \ldots\right) \tag{5.10}$$

as a function of N and T.

The total internal energy of the ideal boson gas is

$$\mathcal{U} = \int_0^{+\infty} E dn = \frac{2\mathcal{Z}k_B T}{\sqrt{\pi}} \int_0^{+\infty} \frac{\zeta^{3/2}}{\exp(\alpha)\exp(\zeta) - 1} d\zeta$$
$$= \frac{3}{2}k_B T \mathcal{Z}\exp(-\alpha)[1 + 2^{-5/2}\exp(-\alpha) + \cdots] \tag{5.11}$$

which, by means of equation (5.10), eventually leads to

$$\mathcal{U} = \frac{3}{2}Nk_B T\left(1 - 2^{-5/2}\frac{N}{\mathcal{Z}} - \ldots\right) \tag{5.12}$$

which represents our main result. Equation (5.12) states that, to the first order, the internal energy of the quantum boson gas is actually given by the classical result already obtained in equation (1.57). Quantum features are just a second order effect, indeed very small: in normal temperature and pressure conditions[7] it is found to be $N/\mathcal{Z} \sim \mathcal{O}(10^{-5})$ for most real gases. In short, to a very good approximation, quantum corrections can be neglected: Boltzmann statistics is just fine. However, since $N/\mathcal{Z} \sim (N/V)T^{-3/2}$, we understand they become important at high number density and low temperature.

5.3 The Bose–Einstein condensation

Boson systems with fixed number of particles undergo a phenomenon referred to as *Bose–Einstein condensation*: intuitively, we understand that by cooling the boson system down to a very low temperature (close to the absolute zero) an increasing number of particles can be accommodated on the ground state (that is on the

---

[6] Just remember that $\alpha > 0$ as discussed in section 3.3 and therefore: $[\exp(\alpha + \zeta) - 1]^{-1} = \exp(-\alpha)$ $[\exp(-\zeta) + \exp(-\alpha - 2\zeta) + \ldots]$.
[7] Namely, at pressure of about 1 atm and room temperature.

minimum energy level) since there is no Pauli principle at work to prevent such a 'condensation'. In this section we aim at providing a simple theory for it, while a more advanced treatment of the Bose–Einstein condensation and related phenomena (like, for instance, superfluidity) is found in more advanced textbooks [5, 6].

We consider a set of N bosons with mass m, confined within a volume V and, for the sake of simplicity, we neglect their mutual interactions. In other words, we fix our attention on *a confined free gas of bosons with fixed number of particles*. We start from equation (3.30) and write

$$n(E,\ T) = \frac{1}{\exp[(E - \mu_{\mathrm{c}})/k_{\mathrm{B}}T] - 1} \tag{5.13}$$

where the α Lagrange multipliers was set similarly to equation (3.15) as discussed in section 3.4.2. We understand that E is the energy of the levels where particles can be accommodated and, in particular, we name E_{GS} the energy of the ground state level. In this boson case we must have $\mu_{\mathrm{c}} < E_{\mathrm{GS}}$ in order to prevent unphysical negative values for the occupation number $n(E,\ T)$. Furthermore, since by decreasing the temperature the number of bosons occupying the ground state increases, then μ_{c} must increase as T drops.

Based on equation (5.13) we now try to predict the occupation number of the ground state as $T \to 0$. We expect a very large value of $n(E_{\mathrm{GS}},\ T \to 0)$ since in this case many particles are accommodated on the ground state, no Pauli principle excluding this accumulation. This implies that the argument of the exponential term is accordingly very small; therefore, we set

$$n(E_{\mathrm{GS}},\ T) \simeq \frac{1}{1 + [(E_{\mathrm{GS}} - \mu_{\mathrm{c}})/k_{\mathrm{B}}T] - 1} = -\frac{k_{\mathrm{B}}T}{\mu'} \tag{5.14}$$

where we have set $\mu' = \mu_{\mathrm{c}} - E_{\mathrm{GS}}$. On the other hand, for any level $E > E_{\mathrm{GS}}$ we get

$$\begin{aligned} n(E,\ T) &= \frac{1}{\exp[(E - \mu_{\mathrm{c}})/k_{\mathrm{B}}T] - 1} \\ &= \frac{1}{\exp[(E - \mu_{\mathrm{c}} \pm E_{\mathrm{GS}})/k_{\mathrm{B}}T] - 1} \\ &= \frac{1}{\exp[(E - E_{\mathrm{GS}})/k_{\mathrm{B}}T]\exp(-\mu'/k_{\mathrm{B}}T) - 1} \end{aligned} \tag{5.15}$$

where the exponential term containing the quantity μ' can be approximated to 1 because of equation (5.14) and by taking into consideration that, as discussed above, $n(E_{\mathrm{GS}},\ T) \gg 1$ in the limit $T \to 0$.

This results allows us to predict the total number N_{ex} of particles occupying excited levels with energy $E > E_{\mathrm{GS}}$ as

$$N_{\mathrm{ex}} = \int_0^{+\infty} \frac{G(E')\mathrm{d}E'}{\exp(E'/k_{\mathrm{B}}T) - 1} \tag{5.16}$$

where $E' = E - E_{GS}$. The density of states for free particles confined in a volume V is given in equation (1.50) and this leads to

$$N_{ex} = 2.612 \left(\frac{2\pi m k_B T}{h^2} \right)^{3/2} V \tag{5.17}$$

where

$$\int_0^{+\infty} \frac{x^{1/2}}{\exp(x) - 1} dx = 1.306 \, \pi^{1/2} \tag{5.18}$$

is a tabulated integral. This result allows us to define the *transition temperature* T_{BE} *for the Bose–Einstein condensation* as that temperature at which all particles are accommodated in excited states. Accordingly, by setting $N_{ex} = N$ in equation (5.17) we obtain

$$T_{BE} = \frac{h^2}{2\pi m k_B} \left(\frac{1}{2.612} \frac{N}{V} \right)^{2/3} \tag{5.19}$$

and below this temperature we calculate

$$\frac{N_{GS}}{N} = \frac{N - N_{ex}}{N} = 1 - \left(\frac{T}{T_{BE}} \right)^{3/2} \tag{5.20}$$

as the *condensation fraction*, that is, the ratio between the number of particles in the ground state N_{GS} and the total number of particles. This fraction varies with temperature, as shown in figure 5.1.

It is important to remark on two features of the Bose–Einstein condensation: first of all, it does not occur in the position space (although it can drive a change in the

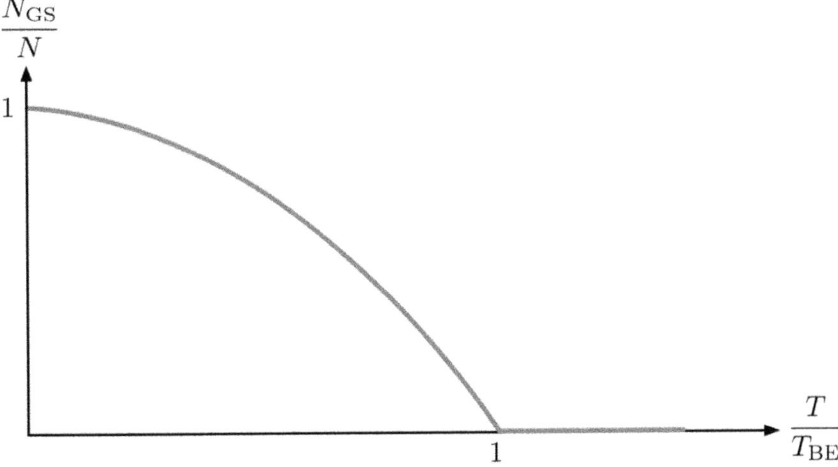

Figure 5.1. The condensation fraction for a system of free bosons confined within a volume. The temperature T_{BE} marks the onset of the transition driving the system towards the Bose–Einstein condensation upon decreasing temperature.

spatial distribution of the particles [5]), but rather in the phase space (since it corresponds to an accumulation of particles on the lowest energy state); next, it is not driven by any kind of physical interaction (actually, we obtained this phenomenon studying a system of free particles), but it is rather dictated by the particle statistics. Although the Bose–Einstein condensation was predicted in 1925 on the basis of theoretical arguments, its first experimental evidence was obtained only in 1995 in a gas of Rb atoms at a temperature as close as 1.7×10^{-7} K.

A real boson system well described by the simple model we developed in this section is liquid ^{4}He. By taking its typical number density, we obtain $T_{\mathrm{BE}}^{^4\mathrm{He}} = 3.13$ K: interestingly enough, this is a temperature very close to 2.17 K, the temperature at which liquid helium is found to undergo a transition to a novel state of matter known as 'superliquid': basically, in the temperature range below 2.17 K ^{4}He behaves like a liquid with vanishingly small viscosity. The similarity between these two temperatures seems to suggest that superfluidity could be looked at as an emerging macroscopic quantum property of a Bose–Einstein condensate: a very intriguing physical situation indeed. The theory we have developed is too simplified to explore this perspective in detail[8] and, therefore, we refer the interested reader to more advanced texts [5, 6].

References

[1] Eisberg R and Resnick R 1985 *Quantum Physics of Atoms, Molecules, Solids, Nuclei, and Particles* 2nd edn (Hoboken, NJ: Wiley)
[2] Demtröder W 2010 *Atoms, Molecules and Photons* (Berlin: Springer)
[3] Cheng T-P 2005 *Relativity, Gravitation and Cosmology* (Oxford: Oxford University Press)
[4] Reif F 1987 *Fundamentals of Statistical and Thermal Physics* (New York: McGraw-Hill)
[5] Kennett M P 2021 *Essential Statistical Mechanics* (Cambridge: Cambridge University Press)
[6] Huang K 1987 *Statistical Mechanics* (New York: Wiley)
[7] Swendsen R H 2012 *An Introduction to Statistical Mechanics and Thermodynamics* (Oxford: Oxford University Press)

---

[8] Let this observation be sufficient: in a sample of liquid helium, atom–atom interactions cannot be entirely neglected—as we did—if one wishes to correctly describe the transition to the superfluid state.

Part III

Concluding remarks

Statistical Physics of Condensed Matter Systems
A primer
Luciano Colombo

Chapter 6

What is missing in this 'Primer'

A 'Primer' cannot be, by definition, a complete and in-depth introduction to any topic. This is especially true for such a huge field as statistical physics. This volume is no exception and, therefore, it must be acknowledged that many important issues have not been dealt with. I feel dutifully committed to mentioning them.

A concise list of missing arguments, as well as of topics which have not been thoroughly treated, is the following:

1. introduction to the probabilistic behaviour;
2. ensemble theory;
3. ergodicity and related issues;
4. kinetic and transport theory;
5. theory of the grand canonical ensemble;
6. phase transitions, order, and critical phenomena;
7. Ising model;
8. non-equilibrium statistical physics;
9. irreversibility;
10. statistical mechanics of complex systems.

Most of the above topics are covered by the textbooks quoted in the bibliography appended to each chapter: for further information, we refer the reader to these manuals. We observe, however, that their level is still introductory, although often more thorough than found in this Primer. The cutting-edge research in statistical physics is further away: at that level, it is preferable to deal directly with the scientific literature published in specialised journals in statistical mechanics.

Part IV

Appendices

Statistical Physics of Condensed Matter Systems
A primer
Luciano Colombo

Appendix A

Mathematical tools

A.1 Stirling formula

The calculation of the quantity $\ln n!$ (where n is a large number) plays a very important role in statistical physics. Let us start by developing the natural logarithm of the factorial number in a very straightforward way

$$\ln n! = \sum_{i=1}^{n} \ln i \tag{A.1}$$

which, of course, holds for integer numbers only. If n is very large, we can to a good approximation replace the discrete sum by an integral

$$\ln n! \simeq \int_{1}^{n} \ln x \; dx \tag{A.2}$$

over a continuous variable x. By using the rule for integration by part $\int u dv = uv - \int v \; du$ and setting $u = \ln x$ and $x = v$ we calculate

$$\int_{1}^{n} \ln x \; dx = [x \ln x - x]_{1}^{n} = n \ln n - n + 1 \simeq n \ln n - n \tag{A.3}$$

since $n \gg 1$. This result is known as Stirling formula.

A.2 Lagrange multipliers

Suppose we need to maximise or minimise a function of n variables $F(x, y, z, \ldots)$ under some given constraints that make the variables not independent. Let us consider for instance the case of two constraints which we cast in the form $\xi(x, y, z, \ldots) = 0$ and $\zeta(x, y, z, \ldots) = 0$.

When the function $F(x, y, z, \ldots)$ is at its maximum or minimum value, its differential with respect to small variations $dx, dy, dz, \ldots$ of its variables is zero

doi:10.1088/978-0-7503-2269-0ch7

$$dF\,(x,\,y,\,z,\,\ldots) = \frac{\partial F}{\partial x}dx + \frac{\partial F}{\partial y}dy + \frac{\partial F}{\partial z}dz + \ldots = 0 \tag{A.4}$$

which, however, does not imply that $\partial F/\partial x = \partial F/\partial y = \partial F/\partial z = \ldots = 0$ since the dx, dy, dz, $\ldots$ are not independent, their variation being restricted as

$$d\xi = \frac{\partial \xi}{\partial x}dx + \frac{\partial \xi}{\partial y}dy + \frac{\partial \xi}{\partial z}dz + \ldots = 0 \quad \text{and} \quad d\zeta = \frac{\partial \zeta}{\partial x}dx + \frac{\partial \zeta}{\partial y}dy + \frac{\partial \zeta}{\partial z}dz + \ldots = 0 \tag{A.5}$$

because of the constraints. In order to eventually solve the $dF = 0$ stationary problem, the above constraints are multiplied by two respective arbitrary constants α and β and then summed to equation (A.4)

$$\left(\frac{\partial F}{\partial x} + \alpha\frac{\partial \xi}{\partial x} + \beta\frac{\partial \zeta}{\partial x}\right)dx + \left(\frac{\partial F}{\partial y} + \alpha\frac{\partial \xi}{\partial y} + \beta\frac{\partial \zeta}{\partial y}\right)dy + \left(\frac{\partial F}{\partial z} + \alpha\frac{\partial \xi}{\partial z} + \beta\frac{\partial \zeta}{\partial z}\right)dz + \ldots = 0 \tag{A.6}$$

where the two parameters are referred to as Lagrange multipliers. The equation (A.6) contains $(n + 2)$ variables, among which the α and β are arbitrary parameters; this therefore implies that

$$\frac{\partial}{\partial x}(F + \alpha\xi + \beta\zeta) = 0 \tag{A.7}$$

and similar equations involving partial derivatives with respect to the other y, z, $\ldots$ variables.

In summary, we derived a set of n relations (as many as the number of variables the function F depends on) of the kind reported in equation (A.7) which allow us to determine the values of x, y, z, $\ldots$ for the critical point of $F(x, y, z, \ldots)$, as function of the α and β parameters. By substituting these values into the constraints $\xi(x, y, z, \ldots) = 0$ and $\zeta(x, y, z, \ldots) = 0$ we straightforwardly obtain α and β.

A.3 Integrals useful in statistical physics

The following integrals are calculated using standard calculus

$$\int_0^{+\infty} \exp\left(-\beta x^2\right) dx = \frac{1}{2}\sqrt{\frac{\pi}{\beta}}$$

$$\int_0^{+\infty} x \exp\left(-\beta x^2\right) dx = \frac{1}{2\beta}$$

$$\int_0^{+\infty} x^2 \exp\left(-\beta x^2\right) dx = \frac{1}{4}\sqrt{\frac{\pi}{\beta^3}} \tag{A.8}$$

$$\int_0^{+\infty} x^3 \exp\left(-\beta x^2\right) dx = \frac{1}{2\beta^2}$$

$$\int_0^{+\infty} x^4 \exp\left(-\beta x^2\right) dx = \frac{3}{8}\sqrt{\frac{\pi}{\beta^5}}$$

IOP Publishing

Statistical Physics of Condensed Matter Systems
A primer
Luciano Colombo

Appendix B

Gibbs entropy

The implicit (section 1.6.2) or explicit (section 1.6.3) assumption underlying the Boltzmann definition of entropy is that in equilibrium all microstates compatible with the same macrostate are equally likely. This is actually not the case for systems out of equilibrium or for states that do not have the same energy (for instance, states defined in the canonical ensemble, characterised by its temperature, volume, and number of particles). We need therefore to generalise this notion.

Let us name π_i the probability of finding the system in the ith microstate, then according to Gibbs we define the system entropy as

$$S_{\text{Gibbs}} = -k_{\text{B}} \sum_i \pi_i \ln \pi_i \tag{B.1}$$

where k_{B} is the usual Boltzmann constant. By using this definition we can develop a full statistical theory, similarly to what has been done in chapter 1, valid even if the condition of the same likelihood for the microstates does not hold. However, this definition appears to be somewhat arbitrary at a first glance; we must reconcile it with our previous knowledge about entropy in order to accept it. We will accomplish this mission by explicitly treating the equilibrium case.

In this condition, two constraints must be obeyed, namely the Gibbs entropy must be maximum and the microstate probabilities must be normalised

$$\sum_i \pi_i = 1 \tag{B.2}$$

which combine into defining the following stationary property for the maximum-entropy condition valid at equilibrium

$$\frac{\partial}{\partial \pi_i} \left[S_{\text{Gibbs}} - \alpha \left(\sum_i \pi_i - 1 \right) \right] = 0 \tag{B.3}$$

doi:10.1088/978-0-7503-2269-0ch8

where α is a Lagrange multiplier (see appendix A). By developing the simple algebra we get

$$-k_B \ln \pi_i - k_B - \alpha = 0 \tag{B.4}$$

a condition which is *just the same for any microstate probability* π_i; this result therefore implies that *in equilibrium all such probabilities must be equal*, as expected according to Boltzmann. If as usual we name the number of equilibrium microstates by Ω, then we immediately get $\pi_i = 1/\Omega$ for any i. Let us now calculate the Gibbs entropy for the equilibrium state

$$\begin{aligned}
S_{\text{Gibbs}} &= -k_B \sum_i \frac{1}{\Omega} \ln \frac{1}{\Omega} \\
&= -k_B \Omega \frac{1}{\Omega} \ln \frac{1}{\Omega} \\
&= k_B \ln \Omega
\end{aligned} \tag{B.5}$$

which corresponds to the Boltzmann entropy defined in equation (1.37).

Statistical Physics of Condensed Matter Systems
A primer
Luciano Colombo

Appendix C

Thermodynamic potentials

C.1 Thermodynamic parameters

We name *internal energy* $\mathcal{U}$ of a thermodynamic system its total energy content; it corresponds to the work needed to form such an aggregate. The internal energy is a *state function* since its value is uniquely defined by the thermodynamic parameters defining the macrostate; we write

$$\mathcal{U} = \mathcal{U}(S,\ V,\ \{n_k\}) \tag{C.1}$$

where $\{n_k\}$ represents in compact form the set of numbers defining how many moles for each $k = 1, 2, 3, \ldots$ different chemical species form the system. Upon differentiation

$$d\mathcal{U} = \left.\frac{\partial \mathcal{U}}{\partial S}\right|_{V,\{n_k\}} dS + \left.\frac{\partial \mathcal{U}}{\partial V}\right|_{S,\{n_k\}} dV + \sum_k \left.\frac{\partial \mathcal{U}}{\partial n_k}\right|_{V,S,\{n_j\}_{j\neq k}} dn_k \tag{C.2}$$

we get the following definitions [1, 2]

$$\text{temperature: } T = \left.\frac{\partial \mathcal{U}}{\partial S}\right|_{V,\{n_k\}}$$

$$\text{pressure: } P = -\left.\frac{\partial \mathcal{U}}{\partial V}\right|_{S,\{n_k\}} \tag{C.3}$$

$$\text{chemical potential: } \mu_{\mathrm{c},k} = \left.\frac{\partial \mathcal{U}}{\partial n_k}\right|_{V,S,\{n_j\}_{j\neq k}}$$

which are known as *intensive parameters* or *state variables*; on the other hand, internal energy, entropy, volume, and mole numbers are *extensive parameters*. Through these definitions we can rewrite equation (C.2) in a more compact and elegant way

doi:10.1088/978-0-7503-2269-0ch9

C-1

$$d\mathcal{U} = T\,dS - P\,dV + \sum_k \mu_{c,k}\,dn_k \tag{C.4}$$

which basically represent an energy balance since the quantities $-PdV$, TdS, and $\mu_{c,k}dn_k$, respectively, represent the mechanical, non-mechanical, and chemical work exchanged during a quasi-static infinitesimal transformation. The exchanged non-mechanical work is referred to as the *exchanged heat* and it is formally written as $dQ = TdS$.

At equilibrium the thermodynamic parameters are such that $\mathcal{U}$ is minimum. Furthermore, they obey restrictions that can be easily obtained by considering the simple case of a monoatomic system made by two sub-systems which we label by '1' and '2'. We consider three different situations where, respectively, the two sub-systems can (i) exchange just heat, or (ii) exchange heat and mechanical work, or (iii) exchange heat and matter. Then, it is easy to prove that in the three cases at equilibrium we have (i) $T_1 = T_2$, or (ii) $T_1 = T_2$ and $P_1 = P_2$, or (iii) $T_1 = T_2$ and $\mu_{c,1} = \mu_{c,2}$, where of course T_i, P_i, and $\mu_{c,i}$ with $i = 1, 2$ are the temperature, pressure, and chemical potential of the two sub-systems. These results prove that the formal definitions provided in equations (C.3) agree with their phenomenological counterparts. Equation (C.1) is the *fundamental thermodynamic equation* since it contains any information about the system. We remark that it is also possible to choose entropy $S = S(\mathcal{U}, V, \{n_k\})$ as the state function of the system, as extensively discussed in chapter 1.

C.2 Thermodynamic potentials

Extensive and intensive parameters have been so far used as independent and derived quantities, respectively. In many circumstances, however, it is more appropriate to use intensive parameters as independent variables (since it is much easier to measure them) and to replace the internal energy by some new *thermodynamic potential*.

Formally, this is done by calculating the *Legendre transformation* $\mathrm{L}_X[\mathcal{U}]$ of the internal energy with respect to a suitable intensive parameter X. Accordingly, we define [1–4]

$$
\begin{aligned}
\text{Helmholtz free energy:} \quad & \mathrm{L}_T[\mathcal{U}] = \mathcal{U} - TS = \mathcal{F}(T, V, \{n_k\}) \\
\text{enthalpy:} \quad & \mathrm{L}_P[\mathcal{U}] = \mathcal{U} + PV = \mathcal{H}(S, P, \{n_k\}) \\
\text{Gibbs free energy:} \quad & \mathrm{L}_{T,P}[\mathcal{U}] = \mathcal{U} - TS + PV = \mathcal{G}(T, P, \{n_k\})
\end{aligned}
\tag{C.5}
$$

representing, respectively, the work exchanged quasi-statically during a transformation occurring at constant temperature (Helmholtz free energy), at constant pressure (enthalpy), and at constant temperature and pressure (Gibbs free energy).

The three thermodynamic potentials allow us to define the equilibrium condition when the system is constrained. For instance, if the system is coupled to a heat reservoir (that is, its temperature is kept constant), then the equilibrium state is that with minimum Helmholtz free energy. On the other hand, if the system is kept at constant pressure (that is, it undergoes the mechanical action of a pressure reservoir) then the equilibrium is defined by the minimum of the enthalpy. Finally, the minimum of the Gibbs free energy defines the equilibrium state of a system coupled to the double action of a heat and a pressure reservoir.

References

[1] Callen H 1985 *Thermodynamics and an Introduction to Thermostatistics* (New York: Wiley)
[2] Dittman R and Zemansky M 1997 *Heat and Thermodynamics* 7th edn (New York: McGraw-Hill)
[3] Goodstein D L 1985 *States of Matter* (Dover: New York)
[4] Kennett M P 2021 *Essential Statistical Mechanics* (Cambridge: Cambridge University Press)

IOP Publishing

Statistical Physics of Condensed Matter Systems
A primer
Luciano Colombo

Appendix D

Calculating the grand partition function of a real gas

As derived in chapter 2 the grand partition function of a real gas with two-body interactions is given by

$$\bar{\mathcal{Z}}_{\text{real}} = \frac{1}{N!}\left[\frac{(2\pi m k_{\text{B}}T)^{3/2}}{h^3}\right]^{N} \int \exp\left(-E_{\text{pot}}/k_{\text{B}}T\right)d\mathbf{R}_1 d\mathbf{R}_2 \cdots d\mathbf{R}_N \qquad (2.4) \quad (\text{D.1})$$

where all symbols are defined in section 2.2.

Let us first evaluate the exponential term

$$\exp\left(-E_{\text{pot}}/k_{\text{B}}T\right) = \prod_{\alpha > \beta} \exp\left[-V(R_{\alpha\beta})/k_{\text{B}}T\right] \tag{D.2}$$

where $R_{\alpha\beta} = |\mathbf{R}_\alpha - \mathbf{R}_\beta|$. By exploiting the twofold fact that (i) $V(R_{\alpha\beta})$ is short-ranged by assumption and (ii) typically $V(R_{\alpha\beta}) \ll k_{\text{B}}T$ we get

$$\exp\left[-V(R_{\alpha\beta})/k_{\text{B}}T\right] = 1 \underbrace{-\frac{V(R_{\alpha\beta})}{k_{\text{B}}T} + \frac{1}{2}\left[\frac{V(R_{\alpha\beta})}{k_{\text{B}}T}\right]^2 + \cdots}_{F_{\alpha\beta}} = 1 + F_{\alpha\beta} \tag{D.3}$$

which allows us to calculate

$$\exp\left(-E_{\text{pot}}/k_{\text{B}}T\right) = \prod_{\alpha > \beta}(1 + F_{\alpha\beta}) \simeq 1 + \sum_{\alpha > \beta} F_{\alpha\beta} \tag{D.4}$$

doi:10.1088/978-0-7503-2269-0ch10

where we have neglected all terms involving products of two or more $F_{\alpha\beta}$'s. For our purposes this is an acceptable approximation, while a more refined theory should include them [1, 2]. The multiple integral can now be calculated

$$\int \exp\left(-E_{\text{pot}}/k_{\text{B}}T\right)d\mathbf{R}_1 d\mathbf{R}_2 \cdots d\mathbf{R}_N = \int \left(1 + \sum_{\alpha>\beta} F_{\alpha\beta}\right)d\mathbf{R}_1 d\mathbf{R}_2 \cdots d\mathbf{R}_N$$

$$= V^N + \frac{1}{2}N(N-1)\int F_{\alpha\beta}d\mathbf{R}_1 d\mathbf{R}_2 \cdots d\mathbf{R}_N \qquad \text{(D.5)}$$

$$= V^N + \frac{1}{2}N(N-1)V^{N-2}\int F_{\alpha\beta}d\mathbf{R}_\alpha d\mathbf{R}_\beta$$

where: (i) the V^N term comes from the the integration of the first contribution appearing in the round parenthesis, assuming that V is the volume of the container the gas is confined in; (ii) the prefactor $\frac{1}{2}N(N-1)$ comes from the fact that all terms in the sum are alike since $F_{\alpha\beta}$ has the very same form for any pair.

In order to evaluate the last remaining integral in equation (D.5) we fully exploit the *central character of the guessed interatomic potential*, defining a two-body problem with spherical symmetry: by placing the origin of a Cartesian frame of reference on atom α we can accordingly set $d\mathbf{R}_\beta = 4\pi r^2 dr$ with $r = R_{\alpha\beta}$. Therefore, we calculate

$$\int F_{\alpha\beta}d\mathbf{R}_\alpha d\mathbf{R}_\beta = \int \underbrace{\left[\int F_{\alpha\beta}4\pi r^2 dr\right]}_{a} d\mathbf{R}_\alpha = \int a \, d\mathbf{R}_\alpha = aV \qquad \text{(D.6)}$$

since the integral a does not depend on the position coordinate $\mathbf{R}_\alpha$. In conclusion, the multiple integral given in equation (D.5) is calculated

$$\int \exp\left(-E_{\text{pot}}/k_{\text{B}}T\right)d\mathbf{R}_1 d\mathbf{R}_2 \cdots d\mathbf{R}_N = V^N + \frac{1}{2}N(N-1)V^{N-2}aV$$

$$= V^N + \frac{1}{2}N(N-1)aV^{N-1}$$

$$= V^N + \frac{1}{2}N^2 aV^{N-1} \qquad \text{(D.7)}$$

$$= V^N\left(1 + \frac{N^2 a}{2V}\right)$$

where we have approximated $N(N-1) \simeq N^2$ with no loss of accuracy since N is typically a very large number (of the order of the Avogadro number).

References

[1] Kennett M P 2021 *Essential Statistical Mechanics* (Cambridge: Cambridge University Press)
[2] Reif F 1987 *Fundamentals of Statistical and Thermal Physics* (New York: McGraw-Hill)

IOP Publishing

Statistical Physics of Condensed Matter Systems
A primer
Luciano Colombo

Appendix E

Fermi–Dirac distribution law: a phenomenological derivation

Let us consider a gas of N free electrons, confined within a volume V in equilibrium at temperature T. Quantum mechanics provides the energy spectrum of this system [1–3] which is discrete and, accordingly, we label by E_i with $i = 1, 2, 3, \ldots$ the energies of the single-electron levels. We ask ourselves *how likely each level is to be occupied*.

Let us assume that two energy levels E_1 and E_2 are occupied with probability $f(E_1, T)$ and $f(E_2, T)$, respectively. Since the system is in equilibrium, on average the number of electrons undergoing the transition $E_1 \rightarrow E_2$ in the unit time equals the number of electrons undergoing the inverse transition $E_2 \rightarrow E_1$; this state of affairs is known as microreversibility[1]. Such transitions could be generated by any kind of mechanism, like electron-electron or electron-defect scattering.

The number of electrons undergoing the transition $E_1 \rightarrow E_2$ per unit time is calculated as the product between the number of electrons initially occupying the level E_1 and the rate $R_{1 \rightarrow 2}$ of occurrence of such transition. The initial number of electrons is in turn given by the product between the total number of particles N and the probability that the initial energy level is occupied. Similar definitions hold for the inverse transition. If electrons were classical particles, the occupation probability of any given level with energy E would be proportional to the Boltzmann factor $\exp(-E/k_\mathrm{B}T)$ as discussed in chapter 1 and, therefore, we could write the *classical version of microreversibility* as

$$\exp(-E_1/k_\mathrm{B}T)\; R_{1 \rightarrow 2}^{\mathrm{classical}} = \exp(-E_2/k_\mathrm{B}T)\; R_{2 \rightarrow 1}^{\mathrm{classical}} \tag{E.1}$$

where $R_{1 \rightarrow 2}^{\mathrm{classical}}$ and $R_{1 \rightarrow 2}^{\mathrm{classical}}$ are the transition rates calculated according to classical physics. Since, however, electrons are identical and indistinguishable quantum

---

[1] Microreversibility is the foundation of the detailed balance principle [4, 5] which is widely used in more advanced formulations of statistical mechanics.

particles they must obey the Pauli principle. This suggests the most general way to switch from classical to quantum transition rates

$$R_{1\to2}^{\text{quantum}} = R_{1\to2}^{\text{classical}} \left[1 - f(E_2, T)\right] \quad R_{2\to1}^{\text{quantum}} = R_{2\to1}^{\text{classical}} \left[1 - f(E_1, T)\right] \quad \text{(E.2)}$$

where the terms in square parenthesis $[\cdots]$ define the probability that the final state is not previously occupied. *Quantum microreversibility* is straightforwardly cast in the form

$$f(E_1, T)\, R_{1\to2}^{\text{quantum}} = f(E_2, T)\, R_{2\to1}^{\text{quantum}} \tag{E.3}$$

which, by inserting equation (E.1), leads to

$$\frac{R_{1\to2}^{\text{classical}}}{R_{2\to1}^{\text{classical}}} = \frac{\exp(-E_2/k_BT)}{\exp(-E_1/k_BT)} = \frac{f(E_2, T)}{f(E_1, T)}\frac{1 - f(E_1, T)}{1 - f(E_2, T)} \tag{E.4}$$

or equivalently

$$\frac{1 - f(E_1, T)}{f(E_1, T)} \exp(-E_1/k_BT) = \frac{1 - f(E_2, T)}{f(E_2, T)} \exp(-E_2/k_BT) \tag{E.5}$$

Since this result applies to each arbitrary pair of energies E_1 and E_2, we logically draw the conclusion that the two expressions appearing on the left and right side of this equation must be equal to the same function, which only depends on temperature. Following the standard convention [4, 6, 7], we name this function $\exp(-\mu_c/k_BT)$ where μ_c is the *chemical potential*. For any arbitrary value of the electron energy E we obtain

$$\frac{1 - f(E, T)}{f(E, T)} \exp(-E/k_BT) = \exp(-\mu_c/k_BT) \tag{E.6}$$

which leads to

$$f(E, T) = \frac{1}{\exp[(E - \mu_c)/k_BT] + 1} \tag{E.7}$$

describing *the probability that the quantum level E is occupied when the electron system is in equilibrium at temperature T.*

The occupation probability defined in this model calculation is the counterpart of the occupation number defined by equation (3.17) of the more general statistical theory leading to the Fermi–Dirac distribution law. In discussing the present electron system any energy level has been considered just one-fold degenerate. In any case, the correspondence between equation (E.7) and equation (3.17) provides evidence that electrons are fermions, consistently with the fact that they are experimentally observed to have spin $\hbar/2$ [8, 9].

References

[1] Colombo L 2021 *Solid State Physics: A Primer* (Bristol: IOP Publishing)
[2] Kittel C 1996 *Introduction to Solid State Physics* 7th edn (Hoboken, NJ: Wiley)
[3] Ashcroft N W and Mermin N D 1976 *Solid State Physics* (London: Holt-Saunders International Editions))
[4] Reif F 1987 *Fundamentals of Statistical and Thermal Physics* (New York: McGraw-Hill)
[5] Lebon G, Jou D and Casas-Vázquez J 2008 *Understanding Non-equilibrium Thermodynamics* (Berlin: Springer)
[6] Glazer M and Wark J 2001 *Statistical Mechanics: A Survival Guide* (Oxford: Oxford University Press)
[7] Swendsen R H 2012 *An Introduction to Statistical Mechanics and Thermodynamics* (Oxford: Oxford University Press)
[8] Colombo L 2019 *Atomic and Molecular Physics: A Primer* (Bristol: IOP Publishing)
[9] Demtröder W 2010 *Atoms, Molecules and Photons* (Berlin: Springer)

IOP Publishing

Appendix F

The conceptual framework for solid-state physics

Ordinary condensed matter systems are made of a very large number of atoms, tightly bound by very strong electromagnetic interactions [1–4]. Below their melting temperature, they show no diffusivity and their physical properties are mostly dictated by the bulk-like atoms, that is, by those atoms that are sufficiently far way from surfaces.

Solids form in rather different ways, among which the crystalline state is of special relevance. Crystals are in fact solid-state systems characterised by translational invariance: as a matter of fact, the space distribution of atoms display long-range order, throughly described by the direct and reciprocal lattices. Periodicity is in fact perturbed by defects or, in other words, real solids are either contaminated or doped as the result of natural or intentional processes, randomly altering their pristine chemistry and/or structural order.

In this Primer we will only focus on the idealised situation (also referred to as 'ideal crystal') of a *perfect solid without surfaces*, as obtained by considering a large defect-free portion of the crystal and by imposing periodic boundary conditions to its surfaces. This simplified picture represents the theoretical framework of minimal complexity catching the most relevant features of solid-state physics: here we put statistical physics to work. More specifically, we will apply the Fermi–Dirac and the Bose–Einstein statistics to some selected fermion and boson systems found in the ideal crystal, respectively, corresponding to the electron and the phonon gas. In the first case, we will address the system of conduction electrons in a metal, while in the second case we will investigate the physics of the lattice vibrational quanta (phonons).

The conceptual framework we adopt to treat physics of a crystalline solid relies on a number of approximations. First of all, electric and magnetic effects, as well as charge currents, are treated according to the classical Maxwell electromagnetism, but for the process of emission or absorption of electromagnetic energy which is described through the concept of a photon [5, 6]; on the other hand, ion and electron physics will be described according to quantum mechanics.

Next, we exploit the basic fact that to a large extent the chemical properties of an atom are dictated by its valence electrons only, while core electrons play a minor role in determining most of the solid-state properties [5, 6]. We therefore assume that a solid is described as a collection of ions and valence electrons. The former, made by the nucleus and the core electrons, are described as point-like objects with a nuclear mass specific of their chemical species and carrying a positive charge.

As a further simplification, we completely neglect magnetic interactions among electrons. This choice is also extended to ions. However, this non-magnetic approximation is not equivalent to disregarding the true existence of the electron spin: rather, electrons (i) follow the Fermi–Dirac statistics and (ii) obey the Pauli principle. Since spin is a relativistic feature [7, 8], neglecting any corresponding coupling term is tantamount to developing a theory in the non-relativistic approximation. Accordingly, we will use everywhere the rest mass m_e for electrons.

Following phenomenological arguments based on the large difference between the ion and electron mass [1–4] we can formally decouple electronic and vibrational degrees of freedom; accordingly, we write the total wave function $\Psi_{tot}(\mathbf{r}, \mathbf{R})$ of the system, depending on the full set of electronic $\{\mathbf{r}\}$ and ionic $\{\mathbf{R}\}$ coordinates, as the product between a total ionic wave function $\Psi_n(\mathbf{R})$ and a total electronic wave function $\Psi_e^{(\mathbf{R})}(\mathbf{r})$. This latter describes the electronic configuration corresponding to the clamped-ion configuration $\{\mathbf{R}\}$. Thanks to this approximation, the electronic and ionic degrees of freedom are separately taken into account.

Although electrons are separately treated from ions, the resulting quantum mechanical many-body problem is still formidably complex. An additional approximation is therefore adopted, consisting in assuming that the full set of electron–ion and electron–electron interactions are represented by a suitable effective one-electron potential. We can equivalently state that, according to this picture, each electron independently moves under the action of a local potential describing its embedding into a system of ions and remaining electrons.

The above approximations are surely severe; nevertheless, they are able to predict quite a few electronic, optical, vibrational, and thermal properties of crystalline solids in pretty good agreement with experiments. By this heuristic argument, we chose to work in this conceptual framework.

References

[1]   Colombo L 2021 *Solid State Physics: A Primer* (Bristol: IOP Publishing)
[2]   Kittel C 1996 *Introduction to Solid State Physics* 7th edn (Hoboken, NJ: Wiley)
[3]   Ashcroft N W and Mermin N D 1976 *Solid State Physics* (London: Holt-Saunders Int. Editions)
[4]   Grosso G and Pastori Parravicini G 2014 *Solid State Physics* 2nd edn (Oxford: Academic)
[5]   Demtröder W 2010 *Atoms, Molecules and Photons* (Berlin: Springer)
[6]   Colombo L 2019 *Atomic and Molecular Physics: A Primer* (Bristol: IOP Publishing)
[7]   Bransden B H and Joachain C J 1983 *Physics of Atoms and Molecules* (Addison-Wesley Longman Limited: Harlow)
[8]   Greiner W 1990 *Relativistic Quantum Mechanics* (Berlin: Springer)

IOP Publishing

Statistical Physics of Condensed Matter Systems
A primer
Luciano Colombo

Appendix G

Bose–Einstein distribution law: a phenomenological derivation

Let us consider a one-dimensional quantum harmonic oscillator; basic quantum mechanics [1–3] provides its energy in the form

$$E_\lambda = \left(\lambda + \frac{1}{2}\right)\hbar\omega \tag{G.1}$$

where $\lambda = 0, 1, 2, 3, \ldots$ is the vibrational quantum number, while ω is its oscillation frequency. The vibrational energy levels are non degenerate and, therefore, according to equation (1.30), the *average energy $\langle E \rangle$ of the quantum oscillator when it is in equilibrium at temperature T* is

$$\langle E \rangle = \frac{1}{\mathcal{Z}} \sum_\lambda \left(\lambda + \frac{1}{2}\right)\hbar\omega \, \exp[-(\lambda + 1/2)\hbar\omega/k_\mathrm{B}T] \tag{G.2}$$

where

$$\mathcal{Z} = \sum_\lambda \exp[-(\lambda + 1/2)\hbar\omega/k_\mathrm{B}T] \tag{G.3}$$

is the partition function. The two series appearing in equations (G.2) and G.3 can be summed after some algebra by using the following results

$$\sum_\lambda \exp(-\lambda x) = \frac{1}{[1 - \exp(-x)]} \qquad \sum_\lambda \lambda \exp(-\lambda x) = \frac{\exp(x)}{[\exp(x) - 1]^2} \tag{G.4}$$

doi:10.1088/978-0-7503-2269-0ch13

provided by standard calculus; eventually we obtain

$$\langle E \rangle = [n(\omega, T) + 1/2]\hbar\omega \tag{G.5}$$

where

$$n(\omega, T) = \frac{1}{\exp(\hbar\omega/k_\mathrm{B}T) - 1} \tag{G.6}$$

represents the average occupation number of the harmonic oscillator with frequency ω.

This result is worthy of note for a twofold reason: (i) on the one side, it confirms the result given in equation (4.27) obtained by formal statistical arguments and (ii) represents a phenomenological derivation of the Bose–Einstein statistics since an harmonic oscillator is, according to quantum mechanics, a zero-spin pseudo-particle.

References

[1] Miller D A B 2008 *Quantum Mechanics for Scientists and Engineers* (New York: Cambridge University Press)

[2] Bransden B H and Joachain C J 2000 *Quantum Mechanics* (Englewood Cliffs, NJ: Prentice-Hall)

[3] Sakurai J J and Napolitano J 2011 *Modern Quantum Mechanics* 2nd edn (Reading, MA: Addison-Wesley)

IOP Publishing

Statistical Physics of Condensed Matter Systems
A primer
Luciano Colombo

Appendix H

Density of states of the blackbody radiation

Let us consider a free particle with mass m confined in a volume V. It is a well known results of elementary quantum mechanics [1–3] that

$$G(E)dE = \frac{V}{4\pi^2\hbar^3}(2m)^{3/2}E^{1/2}dE \tag{H.1}$$

represents the number of allowed quantum states per unit energy interval at the energy E. Since the particle is free, then $E = p^2/2m$, where p is its momentum and therefore

$$G(p)dp = \frac{V}{2\pi^2\hbar^3}p^2\,dp \tag{H.2}$$

represents the number of allowed quantum states with corresponding momentum in the interval $[p, p + dp]$.

Blackbody radiation is a transverse electromagnetic field trapped within a cavity dug into a solid body in equilibrium at temperature T. Each radiation mode of frequency ν consists of photons with energy $E = h\nu$. Their momentum $p = h/\lambda$ is provided by the de Broglie relation [4, 5], where $\lambda = c/\nu$ is the mode (or photon) wavelength. In conclusion, by means of the de Broglie relation we can use equation (H.2) as well for photons so that

$$G(\nu)\,d\nu = 2\frac{4\pi V}{c^3}\nu^2\,d\nu \tag{H.3}$$

where the prefactor 2 has been added since there are two independent directions of (transverse) polarisation for each frequency. Equation (H.3) represents the number of states in the blackbody radiation with frequency in the interval $[\nu, \nu + d\nu]$.

doi:10.1088/978-0-7503-2269-0ch14

References

[1] Miller D A B 2008 *Quantum Mechanics for Scientists and Engineers* (New York: Cambridge University Press)

[2] Griffiths D J and Schroeter D F 2018 *Introduction to Quantum Mechanics* 3rd edn (Cambridge: Cambridge University Press)

[3] Bransden B H and Joachain C J 2000 *Quantum Mechanics* (Englewood Cliffs, NJ: Prentice-Hall)

[4] Eisberg R and Resnick R 1985 *Quantum Physics of Atoms, Molecules, Solids, Nuclei, and Particles* 2nd edn (Hoboken, NJ: Wiley)

[5] Demtröder W 2010 *Atoms, Molecules and Photons* (Berlin: Springer)

www.ingramcontent.com/pod-product-compliance
Ingram Content Group UK Ltd.
Pitfield, Milton Keynes, MK11 3LW, UK
UKHW051941150726
7214IPUK00020B/365